CREWE WORKS
∽ A CELEBRATION OF STEAM ∽

KEITH LANGSTON

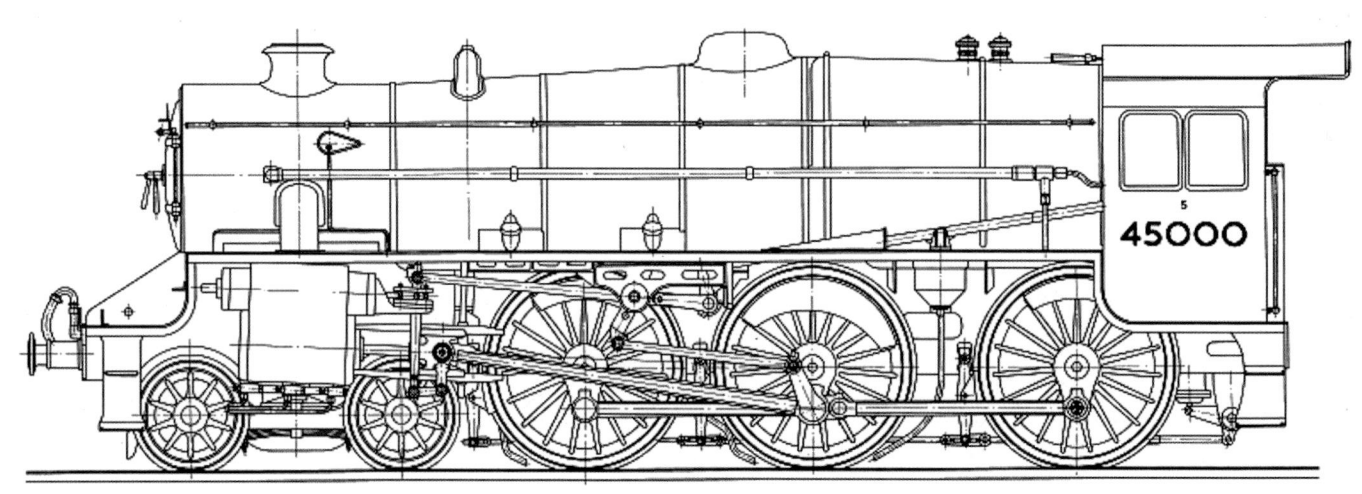

Published in Great Britain by Gesley Books
An imprint of Mortons Books Ltd.
Media Centre
Morton Way
Horncastle LN9 6JR
www.mortonsbooks.co.uk

© 2024 by Gresley Books

All rights reserved. No part of this publication may be reproduced or transmitted in any form or by any means, electronic or mechanical including photocopying, recording, or any information storage retrieval system without prior permission in writing from the publisher.
ISBN 978-1-911704-18-8

The right of Keith Langston to be identified as the author of this work has been asserted in accordance with the Copyright, Designs and Patents Act 1988.

Typeset by Druck Media Pvt.

Contents

1	Made in Crewe
3	Introduction
6	The Works
7	Locomotive Engineering – Images from the Archives
21	Maintenance and Repair
34	Employees
35	LMS Stanier '3MT' 2-6-2T
36	LMS & SDJR Fowler '2P' 4-4-0
41	LMS Ivatt '2MT' 2-6-2T
51	Crewe Belt System
52	LMS Hughes/Fowler 5MT 'Crab' 2-6-0
57	LMS Stanier '5MT' 2-6-0
63	LMS & SDJR Fowler '4F' 0-6-0
67	LMS Stanier 'Black Five' 5MT 4-6-0
77	LMS Fowler 'Patriot' 5XP, 6P5F, 7P 4-6-0
83	LMS Stanier 'Jubilee' 5XP, 6P, 7P 4-6-0
89	LNWR Bowen Cooke 'Claughton' 5XP 4-6-0
93	Climbing Beattock
95	LMS Stanier 'Princess Royal' 7P, 8P, 4-6-2, (Inc. Turbomotive)
101	Stanier Streamlined Pacifics
102	LMS Stanier 'Princess Coronation' or 'Duchess' 7P, 8P, 4-6-2
110	The Coronation Scot 1937 – 1939
117	LMS Ivatt '2MT' 2-6-0
123	LNWR Webb '1P' 2-4-2T
124	LNWR Webb '2P' 0-6-2T. LNWR Webb Bissel Truck '1F' 0-4-2PT
125	LNWR Bowen Cooke '6F' 0-8-2T
126	LNWR Beames '7F' 0-8-4T
127	LMS/War Department (WD) Stanier '8F' 2-8-0
135	LNWR Whale '19in Goods' 4-6-0
136	LNWR 'G1' 6F & 'G2A' 7F 0-8-0
141	LNWR 'G2' 7F 0-8-0
143	LMS Fowler '7F' 0-8-0
144	LNWR Bowen Cooke 'Prince of Wales' 4P 4-6-0
145	LNWR Whale 'Precursor' 3P 4-4-0
146	LNWR Bowen Cooke 'George the Fifth' 3P 4-4-0
147	LNWR Webb 'Chopper Tanks '1P 2-4-0T. LNWR Webb 'Coal Engine' 2F 0-6-0
149	LNWR Webb '18in Goods' 2F 0-6-0. LNWR Webb 'Saddle Tank' 2F 0-6-0ST
150	LNWR Webb 'Special Tank' 2F 0-6-0ST. LNWR Whale 'Rail Motor' 0-4-0T
151	LNWR Webb 'Coal Tank' 0-6-2T
153	British Railways 'Standard Design Group'
155	BR Standard 'Britannia' 7P6F 4-6-2
165	BR Standard 'Duke of Gloucester' 8P 4-6-2
167	BR Standard 'Clan' 6P5F 4-6-2
172	BR Standard 'Class 2 Tank' 2MT 2-6-2T
175	BR Standard '9F' 2-10-0
181	BR Standard Franco-Crosti 2-10-0
187	The Great Gathering
198	Lament to the End of Steam – Bibliography – Acknowledgements

Stanier Coronation Pacific BR No 46220 CORONATION is seen prior to departing Edinburgh Princess Street station with a Glasgow Central local service, circa 1960. David Anderson.

Preserved LNWR Super D 'G2' 0-8-0 BR NO 49395, seen at Churnet Valley Railway during post rebuild 'running in' trials during the summer of 2006. David Gibson.

MADE IN CREWE

Oiling around. Keith Langston.

Steam locomotive building at Crewe Works was carried out by four railway companies, The Grand Junction Railway from 1843, The London & North Western Railway from 1846, The London Midland & Scottish Railway from 1923 and lastly British Railways from 1948. The original 7 acre triangular site built to the north of the station soon became too small as locomotive building and repair increased and over time the works and its workforce were steadily expanded. Well over 100,000 employees have passed through the 'works' since 1843. A total of 100 individual steam locomotive classes were built totalling over 7,330 locomotives, additionally in excess of 125,000 locomotive general repairs were carried out.

The timeline of this publication highlights the 38 steam locomotive classes which were listed as being built by Crewe Works (in any time period) but importantly included in the BR January 1948 stock list. It also includes BR Standard class locomotives built at Crewe post 1951.

Many trades and skills were employed during the building of steam locomotives. Two scenes from the British Railways era. Crewe Works Archive.

Introduction

Crewe Works was founded by the Great Junction Railway (GJR) in 1843 and in 1845 amalgamated with the Liverpool & Manchester Railway. In 1846 that entity amalgamated with the London & Birmingham Railway to form the London North Western Railway (LNWR). Following the 1921 Railway Act (historically referred to as 'Grouping') the Cheshire based works was incorporated into the London Midland & Scottish Railway (LMS) in 1923. By 1947 Crewe Works was responsible for the maintenance of some 2,779 steam locomotives in addition to carrying out a comprehensive locomotive building programme.

By a further act of parliament (The Transport Act 1947), which came into force at midnight on 31 December 1947 the aforementioned four companies, and several other smaller concerns, were 'Nationalised' forming the new entity British Railways (BR), of which Crewe Works was an important part. When the Grand Junction Railway commenced building the works circa 1833 none other than the celebrated railway engineer *George Stephenson* was appointed as 'Engineer in Charge'.

Built at Crewe, one of Stanier's finest.

- Railway builder *Joseph Locke* who replaced Stephenson as 'Engineer in Charge' in 1835.

- In 1840 *William Barber Buddicom* was appointed 'Locomotive Superintendent' but resigned just over a year later and was replaced by *Francis Trevithick*.

- In 1857 *John Ramsbottom* was appointed as LNWR 'Northern Division Locomotive Superintendent' at Crewe. In 1862 Ramsbottom was appointed 'Locomotive Superintendent' for the entire LNWR.

- In September of 1871 Ramsbottom resigned, the official reason being ill health, and he was replaced by *Francis William Webb* from 1 October. But ill health (also described in some records as a breakdown) may not have been the sole reason for his departure. Reportedly he had failed in his effort to negotiate a higher salary with the company directors?

- After Ramsbottom's departure the LNWR continued to pay him an 'annual stipend' in recognition of his usefulness as a consultant. In 1883, he became a consulting engineer and a director of the Lancashire & Yorkshire Railway, for whom he established Horwich Works on a site near Bolton. It was not until 26 years after his 'ill health' retirement that John Ramsbottom died, in 1897 aged 82!

- In May 1903 F.W. Webb also retired with ill health and was replaced by *George Whale* as Chief Mechanical Engineer, on 1 July. But therein hangs a sad tale. Although historic accounts vary in slight detail, it seemed likely that Mr Webb suffered a very serious mental breakdown. Details of that sad happening can be found in the late Brian Reed's excellent book 'Crewe Locomotive Works and its Men', a David & Charles publication.

- George Whale retired in 1909 and was succeeded by *C.J. Bowen Cooke* from 1 July 1909. He was in charge during the period of WWI (1914-1918) when the works was heavily involved in the production of munitions. In the calendar year 1918-1919 he also served as the mayor of Crewe. In 1920 Bowen Cooke, who was known to be suffering an ongoing illness consequently died 'in service' (18/10/1920). He was replaced as Chief Mechanical Engineer by *H.P.M. Beames* from 1 November 1920.

- In 1922 the LNWR merged with the Lancashire & Yorkshire Railway and *George Hughes* of the L&Y was appointed CME of the whole system.

- Hughes' tenure was a short one as he was replaced as CME (following the 1January 1923 creation of the London Midland & Scottish Railway) by H.P.M. Beames.

- *Sir Henry Fowler became* CME in 1925 and served until 1930.

- He was succeeded by *E.J.H. Lemon* in 1931.

- In 1932 the man almost everyone readily associates with Crewe Works arrived! *William A. Stanier* was recruited, or as some historians prefer to say 'Head Hunted' from the Swindon Works of the Great Western Railway (GWR). He served as CME until his retirement as *Sir William A. Stanier F.R.S.* in 1944. His name is associated not only with the complete reorganisation of the works but with the introduction of many locomotive types which were to achieve iconic status. It is said that Stanier brought with him from the GWR a chest containing a large number of detailed working drawings!
But his greatest asset was 40 years of locomotive design and operational experience. Stanier designed or converted 14 LMS locomotive classes comprising 2,431 engines.

- In addition to the steam locomotives built during Stanier's era it should be noted that Crewe Works built 161 'Covenanter Tanks' which were designed at Derby, and also manufactured parts for large artillery pieces.

- After Stanier's retirement *C.E. Fairburn* served as CME until 1945.

- *H.G. Ivatt,* a man with an established reputation as a talented locomotive engineer and designer became CME in 1945 and served until 1951.The man whose name became synonymous with the BR Standard locomotive designs was *R.A. Riddles.* A real 'product' of Crewe in particular, and the LMS in general *Robert Arthur Riddles* joined the LNWR as a premium apprentice in 1913. He served with the 'Royal Engineers' mainly in France during the 1914-18 Great War, during which time he suffered serious injury.

- After the war Riddles returned to Crewe and in 1920 was tasked with overseeing the progress of the new 'Erecting Shop'. His steam engineering knowledge was further enhanced by serving for a spell at Derby Works, where he enjoyed the tutelage of H.G. Ivatt. During the General Strike of May 1926, he volunteered as a driver and stated that he 'gained valuable first-hand knowledge of steam locomotive operating'. In 1939 (WWII had just started) he moved to Ministry of Supply as 'Director of Transportation Equipment' and later designed the WD Austerity both the 2-8-0 and 2-10-0 locomotives. In 1943 he returned to Crewe Works.

- Upon the creation of the Railway Executive in 1947, and prior to 1948 the nationalisation of the railways, he was appointed 'Member of the Railway Executive for Mechanical and Electrical Engineering'. He had two principal assistants, both of whom were also former LMS men, *Roland C. Bond*, Chief Officer (Locomotive Construction and Maintenance), and *E. S. Cox*, Executive Officer (Design). The duties of these three effectively covered the old post of Chief Mechanical Engineer; they oversaw the design of the British Railways (BR) Standard classes.
Mr Riddles retired from BR in 1953 and was succeeded as CME by *Roland Bond*.

The names of many of the 'Crewe Greats' are rightly included in the class titles of numerous steam locomotives in general, and in particular to those 'Built at Crewe'.

Under the LMS and BR the works was closely linked to the former Midland Railway (MR) Derby Locomotive Works and also the former Lancashire & Yorkshire Railway (L&Y) Horwich Locomotive Works.
The individual listings will illustrate that several steam locomotive classes were built jointly and/or concurrently at Crewe, Derby and Horwich works. Those included the LMS Fowler 'Patriot' 4-6-0 engines, LMS Stanier 'Jubilee' 4-6-0 engines, and the LMS Hughes/Fowler Horwich 'Crab' 2-6-0 engines.

1941 Crewe built Stanier/WD '8F' 2-8-0 BR No 48130. Keith Langston Collection.

In order to meet the demands of the WWII locomotive building programme the Stanier '8F Class' 2-8-0 locomotives were not only built at Crewe Works but also at Ashford Works, Brighton Works, Darlington Works, Doncaster Works, Eastleigh Works, Horwich Works and Swindon Works. Additional engines were built by contractors, namely Beyer Peacock & Co Ltd, North British Locomotive Co Ltd and Vulcan Foundry Co Ltd. A truly integrated locomotive construction programme!

At the end of the war, and upon the formation of BR the stock of Stanier '8F' 2-8-0 locomotives were categorised thus, LNER engines built 1943/45, GWR engines built 1943/45, engines built for the LMS, SR engines built 1942/44 and War Department engines built by the aforementioned contractors from 1940. The engines all officially lost their wartime numbers in 1948 and were allocated BR numbers 48000 – 48775. However, one LNER engine did not return to the LMS before 1 January 1948. Doncaster built LNER No 3554, designated 'O6' Class, BR No 63554 which returned to BR after 1 January 1948, and was then renumbered 48759. From January 1948 onwards the re-numbering of the former LMS engines commenced with the addition of 40000 to their numbers, filling the series 40001 to 59999.

Crewe also shared the building of other notable locomotive classes. For example, the Fowler designed '4F' 0-6-0 locomotives were jointly built at Derby Works, St. Rollox Works and by the contractors Sir W.G. Armstrong Whitworth & Co Ltd, North British Locomotive Co Ltd and Kerr Stuart & Co Ltd. The building of the ubiquitous Stanier 'Black Five' 5MT 4-6-0 locomotives was also carried out at other locomotive building centres namely Derby Works, Horwich Works, and contractors Sir W.G. Armstrong Whitworth & Co Ltd and Vulcan Foundry Co Ltd. During the 'BR Standard' locomotive building programme Crewe Works shared construction of the 'Class 2' 2-6-2T engines with Darlington Works and the '9F Class' 2-10-0 engines with Swindon Works. The Standard '9F' class of freight engines heralded the end of steam locomotive production in Great Britain.

Stanier 'Black Five' 5MT 4-6-0 BR No 44969 is seen near Elvanfoot, whilst heading south on the West Coast Main Line (WCML) in 1959. David Anderson.

The Works

Crewe Locomotive Works was originally built for the Grand Junction Railway (GJR) in 1843 and became part of the London North Western Railway (LNWR) in 1846. The site chosen was in the fork formed by the main line to Liverpool and Scotland and the branch to Chester and Holyhead, it was situated to the north of Crewe railway station. The facility was further extended in 1862 and again in 1864. Crewe Locomotive Works was as a result unique as the largest company owned works in the world. During its first few years of operation the works reportedly employed 161 men, but over the following one hundred years the workforce grew to become approximately 8,587 persons by 1943 (employment reportedly peaked higher at approximately 10,000 during times of maximum production). It is true to say that the works and the town grew together and many close links between the two entities were created. The civic dignities of Crewe often included prominent works personnel amongst their number, and the reverse was also commonplace.

The works developed to become self-sufficient in satisfying the majority of its requirements. Its own 'Bessemer' steel plant was opened in 1864 and developed along with the appropriate foundries. A gas making plant was commissioned in 1886, and around that time the LNWR enhanced the capacity of its works water supply (reservoirs) in order to also service the needs of the townsfolk. To assist in transporting materials around, and within the works an 18-inch gauge railway was constructed. In December 1866 the first real milestone was reached as the 1000th steam locomotive was completed, it was a 'DX Class' 0-6-0 No 613.

'Modern Gas Producing Plant' circa 1926. Crewe Archive

Locomotive building continued at a healthy pace, accordingly in 1870 the works was again extended.
In August 1876 the 2000th locomotive was completed, it was a Webb 2-4-0T No 2233.
A 2-2-2-2 'Compound' engine No 600 was completed in June 1887, becoming the 3000th steam locomotive to be built at Crewe Works. In March 1900 the 4000th locomotive was constructed, that being 'Jubilee Class' 4-4-0 No 1926 LA FRANCE. The 5000th locomotive milestone was passed in June 1911 when 'George the Fifth' class 4-4-0 No 5000 CORONATION was completed.

LNWR C. J. Bowen Cooke designed 'George the Fifth' Class 4-4-0 No 5000 CORONATION. E. Talbot Collection.

On 21st April 1913 Crewe Works received a Royal visit by King George V and Queen Mary.
The official 6000th steam locomotive built was LMS 'Crab' 2-6-0 No 13178, completed in 1930.
In June 1937 Stanier's first 'Princess Coronation Class' 4-6-2 LMS No 6220 CORONATION, left the works amid great excitement and much anticipation. The first of the '8P' Stanier Pacifics appeared in LMS 'Streamline' form, ostensibly to meet the challenge of Gresley's streamlined pacific 'A4' class on the LNER.
1950 saw the 7000th locomotive built when in September 'Ivatt 'Class 2' 2-6-2T No 41272 was completed.
1948 was the year in which British Railways (BR) was created.
In 1951 the first BR standard locomotive was completed at Crewe Works, Pacific No 70000 BRITANNIA.
1958 was the year when the last steam locomotive was completed at Crewe Works when in December '9F' 2-10-0 BR No 92250 rolled out of the works. It became officially the 7,331st locomotive to be built at Crewe.

Locomotive Engineering - Images from the Archives.

The spectacular wonder of watching metal run liquid, and seeing sparks fly was once all in a days work for the men of Crewe's steelworks. This image is dated circa 1925. The melting point of iron alloys and steel is approximately 2,200 – 2,500 °F (1,250 – 1370°C). It is hard to believe that the once bustling Crewe Steel Works foundry is in the 21st century the site of a supermarket, where in modern times the descendants/relatives of those who once melted the metals may perchance purchase their groceries! Note the plaque on the steel beam showing the then 'New Foundry' opening date, R. Moon Esq (later Sir Richard Moon) was a director of the LNWR 1861-1891. Steel production at Crewe ceased in 1932.
BR London Midland Region.

An older image of metal production in another area of the steel works (LMS circa 1930). This gentleman would not have reason to complain of being cold at work. The furnace is tapped, and the molten metal runs into the crucible which the overhead lifting gear with heavy hook and chain will later lift, so that the 'hot' vessel can be lifted onto a wagon, in order to be transported to where it is needed, by the works narrow gauge railway. The probable destination would be a foundry area where the metal will then be poured into a specially prepared mould/pattern, for example during the manufacture of a locomotive cylinder.
Image taken prior to 1930. Crewe Works Archive.

Team work. The crucible can be seen lifted from the railway truck, which has brought it from the furnace area to the foundry area. The gentleman to the right is turning a ratcheted handle in order to allow the molten metal (probably iron) to be poured into the moulds which are located in the 'sand bed' on the foundry floor.
The gentleman to the far left is watching, and signalling the rate of flow. Importantly the foundryman to the left rear is using a special long paddle in order to stop the scum (dross) floating on the molten metals surface, from entering the moulds. *John Pritchett Collection.*

More team work, this time 'putting the squeeze on' Crewe Works style. Although it appears that they who watch also serve, the gentleman far right in the bowler hat ('blocker') is in charge, as the foreman. This huge hydraulic press was capable of applying pressures up to the equivalent to 2000 ton per square inch to metal placed in its jaw. The work piece is held whilst balanced on the overhead lifting equipment, and manipulated by the 'gang' (customary flat caps) holding the former and their 'chargehand' (handkerchief on head) can also be seen. The two gentlemen to the far left have control of the press. *Crewe Works Archive.*

The Iron Foundry, note the plethora of moulds, patterns and core bixes on the sand floor area. This image was taken circa December 1899. Crewe Works Archive.

The Brass Foundry. Amongst the foundry workers we have a gentleman in a bowler hat in this posed image. Crewe Works Archive.

Locomotive frames being cut with a guide mounted acetylene torch during the LMS era. Crewe Works Archive.

A 4-cylinder locomotive under construction. Amongst the Crewe built 4-cylinder engines was the Webb 'B' Class 1901-1904, the Bowen Cooke 'Rebuilt Claughton' engines 1913-1921 and the iconic Stanier 4-cylinder 'Princess Royal' and 'Princess Coronation or Duchess' classes 1933-1948. Crewe Works Archive.

A view of the boiler shop seen circa January 1890. In the centre row there can be seen boiler and firebox assemblies for the smaller Webb classes of engines. To the left are a line of larger engines, possibly part of the construction of 8-coupled locomotives. The image was taken on a non-working day but when in full scale production this facility would have been a very noisy work area. Crewe Works Archive.

A completed boiler and Belpaire firebox with dome, but at that time without a smoke box and chimney. It is seen circa 1900 posed outside the boiler shop. This boiler barrel has three sections, and in this case the steam dome is located on top of the centre section.
Note the large amount of riveting work which has taken place. Also the sloping bottom/floor of the ash pan, which is situated below the grate area/firebars. Crewe Works Archive.

The Belpaire firebox was invented by Belgium engineer Alfred Belpaire in 1864. The design was adopted by most British locomotive manufacturers, but notably not by the LNER who preferred the round topped firebox for their locomotives.
The main advantage of the Belpaire design is that it has a greater surface area at the top of the firebox, where most heat is generated. Compared to the round topped firebox the Belpaire shape was more difficult to manufacture, but British locomotive builders were firmly of the opinion that the advantages in steam raising far outweighed the disadvantage of higher cost.
Crewe Works developed a system incorporating a machine, which formed the firebox wrapper into the required shape in one operation.
In the diagram the dotted line represents the boiler and 'S' signifies the steam generating area, where the firebox stays can be seen. Below that area the rear tube plate and tubes can be seen. 'B A' represents the position of the 'Brick Arch' above the fire, and 'G' the grate and firebars.

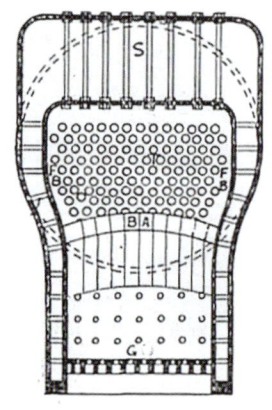

Original drawing by J.F.Gairns.

A fitters platform at use in the 'Erecting Shop'. The right height and therefore safer for the men, and also a shelf located behind them for tools and components. Crewe Works Archive.

Locomotives under construction in the 'New Erecting Shop'. Crewe Works Archive.

In this 1910 view of the erecting shop the foreman, bowler hat and highly polished shoes is supervising work on a Bowen Cooke 4-4-0 locomotive. *John Pritchett Collection/E.Talbot Collection.*

Vertical 'Wheel Lathes' in use, note the belt drives. Crewe Works Archive

Wheel Balancing Machine.

Conn-Rod boring machine.

Both images Crewe Works Archive

Shrinking tyres (steel hoops) onto cast locomotive wheels using directly applied flames. Crewe Works Archive.

View of a machine shop, circa 1906. Crewe Works Archive.

Milling Machine capable of being loaded with several workpieces at the same time, is seen in operation, circa 1930. Crewe Works Archive.

The manufacture of locomotive boiler tubes is seen in this image. Crewe Works Archive.

The firebox, boiler barrel and sloping top smokebox of 'Stanier Pacific' LMS No 6220 CORONATION is seen lifted prior to being united with the frames and rolling chassis, in this 1933 image. *Richard Metcalf Collection*

A Stanier LMS 'Princess Coronation or Duchess Class' 8P Pacific locomotive is seen under construction. The streamlined cladding (casing) is in the process of being fixed into place. It is not hard to see why critics of the locomotives streamlined casing used the mildly derogatory term 'Upturned Bath Tub' to describe it! *Richard Metcalf Collection*

BR 'Standard Pacific' No 70000 BRITANNIA seen at the works in 1950. Work to build the first BR Standard was coming to an end.
Crewe Works Archive.

The rolling chassis of preserved No 70000 is seen at LNWR Heritage during a refurbishment in 2010.
Keith Langston.

BR Caprotti 'Standard Pacific' No 71000 DUKE OF GLOUCESTER visited the works for a repaint during 2004. The iconic locomotive is seen inside the paint shop whilst in primer. Keith Langston with the kind permission of Bombardier Transportation.

Maintenance and Repair.

Although the building of locomotives in general, and in particular express passenger locomotives are often uppermost in the minds of railway enthusiasts, it needs to be noted that the most important functions of a steam locomotive works is repair and maintenance. From the day that the first fire is placed in the firebox of a steam engine boiler the process of degradation has begun. Routine visits to the works for heavy repairs were planned into the locomotives schedule.

Some items of maintenance would be carried out at the locomotives running shed, e.g. re-turning the tyres, refitting axle boxes and bearings, repairs to the motion and associated brake gear and the fitting of new springs.

Locomotive repairs and maintenance continued alongside 'new built' work.
Crewe Works Archive.

Operating pressure vessels, manufactured from ferrous materials to facilitate the high temperature combustion of bituminous coal in order to convert water into pressurised steam, is not a process without problems. Overtime the boiler and its constituent parts will degrade and start to lose the efficiency needed to produce the correct volume of steam. Additionally consider that the whole steam raising ensemble is travelling, often at speed, on a rivetted/welded steel frame holding cylinders connected to reciprocating motion. Huge stresses are created and no matter how skilfully the locomotive has been assembled, and then handled in traffic, failures did occur necessitating the locomotive being taken out of traffic.

A BR 'heavy repair' involved a trip to the works. The requirements of a heavy repair meant either that the engine had to be reboilered, or the boiler be taken out of the frames for repairs. It also included any two of the following requirements :-
- New tyres fitted to 4 or more wheels.
- The fitting of new cylinders.
- Fitting new axle(s) to the engine and/or tender.
- Boiler re-tubing.
- Turning-up wheels and fitting new axle boxes, or motion and brake work overhauled.
- Boiler repaired in the frames, with not less than 50 stays renewed.

An intermediate or 'light repair' was dictated by the combined factors of mileage run and time in traffic. It could consist of any of the requirements stated but which did not separately or together indicate the need for a 'heavy repair'. The work load figures of 'erecting shop south' for the year 1952 are worthy of note. In that year the facility handled 966 'heavy repairs', 625 'intermediate repairs', and additionally built 41 new steam locomotives.

After the railways were Nationalised in January 1948 the newly created British Railways (BR) listed over 2,500 steam locomotives as being Crewe Works allocated for general repairs. That number included 49 Pacifics, 262 3-cylinder 4-6-0s, 636 'Class 5' 4-6-0s, 535 'Class 8' 2-8-0s, 218 ex LMSR 0-6-0s, 218 'Class 4' 2-6-2Ts, and over 500 ex-LNWR and ex-LMS engines from up to 12 separate classes. New locomotive construction was also still under way and included, Stanier/Ivatt Pacific BR No 46257 CITY OF SALFORD, approximately 50 'Class 5' 4-6-0s of several variations and a variety of Ivatt small 2-6-0s and 2-6-2Ts. The 'British Railways Standard Locomotive' building programme delivered its first locomotive from Crewe on 2 January 1951, when BR No 70000 BRITANNIA steamed out of the works. As stated earlier the last steam locomotive to leave the works was BR Standard '9F' 2-10-0 No 92250 in December 1958.

Steam locomotive repairs and general maintenance continued, even as the new era of diesel and electric locomotives began at the works. The last steam heavy repair was to 'Britannia Pacific' No 70013 OLIVER CROMWELL, completed on 2 February 1967. That repair was said to mark approximately 125,000 steam locomotive repairs at the site since 1843. As the date for the end of steam traction on BR 11 August 1968 had already been set in stone it was remarkable that No 70013 was repaired at all. Perchance the decision already made to preserve the 'Britannia Class' Pacific as part of the National Collection, and indeed to later use it extensively on 'end of steam' charters, justified the locomotives final trip to the works!

'Patriot Class' 4-6-0 BR No 45516 THE BEDFORDSHIRE AND HERTFORDSHIRE REGIMENT, pictured outside the erecting shop at Crewe Works in January 1949. *Edward Talbot Collection.*

National Collection BR 'Standard Britannia' 4-6-2 No 70013 OLIVER CROMWELL is seen passing Rose Grove on 21/07/68.
The occasion was one of the 'Roch Valley Railway Society' Manchester-Southport steam excursions on 21/07/68, reporting No 1T40.
BR No 70013 was withdrawn from Carnforth (10A) during the week ending 17/08/1968, and preserved.
Michael Halbert Collection.

The use of ear protection seemed not to be a requirement during pre-millennial railway engineering operations, as this Crewe Works image of a gentleman using a 'Pneumatic Stud Driver' shows. Crewe Works Archive.

A Horizontal Wheel Lathe in use. Perchance a modern day Health & Safety nightmare. Note the circular pockets around the circumference of the wheel boss, they were to allow for the fitting of balance weights. Crewe Works Archive.

View of the Iron Foundry dating from 30 December 1899. Two 'Jib' (or Pivot cranes) can be identified in the centre of the image. Also two overhead cranes can be seen spanning the work area, they were operated from the floor by chains. The usual assortment of moulds, patterns and core boxes are in evidence. On the third upright to the left and also on the Jib Crane are sieves, for use when moulding sand. Crewe Works Archive.

Simply put, with oxy acetylene (oxy-fuel) cutting, a torch is used to heat metal until it attains a red glow thus indicating that it has reached its 'kindling' temperature. A stream of pressurised oxygen is then trained on the metal, burning it into a metal oxide that flows out of the kerf (groove or slit) as dross (waste). This process, in some form or other has been in use for well over 125 years. Crewe Works Archive.

The 'Machine Shop Arcade', note the belt drives and mostly unguarded machinery. Seen circa 1930. Crewe Works Archive.

BR Standard 'Britannia' Pacific No 70011 HOTSPUR, the locomotive is seen whilst undergoing its last overhaul, on 16/01/1966. Rail Photoprints Collection/Brian Robbins.

In addition to repairing and maintaining locomotives the works also scrapped engines and often recycled the metal. Two former North Staffs Railway (NSR) 0-4-4T's await scrapping outside the foundry at Crewe Works, on 07/05/1939. Nearest the camera is Adams designed 'Class 15' LMS No 1436 built 1920 and behind it Adams designed 'Class 9' LMS No 1434 built 1908, alongside is a Webb Coal Tank which also awaits its fate along with two Claughtons. George C. Lander/www.railphotoprints.co.uk.

BR Standard 'Britannia' Class Pacific No 70007 COEUR-DE-LION is seen at Crewe whilst still in Grey Primer on 18/03/1951. The completed locomotive left the works a month later and was withdrawn from Carlisle Kingmoor (12A) in the week ending 19/06/1965. Mike Bentley Collection.

Stanier 'Princess Royal' Class Pacific LMS No 6201 PRINCESS ELIZABETH is seen during a works visit circa October 1947. This locomotive was withdrawn from Carlisle Upperby (12B) during the week ending 20/10/1962. Rail Photoprints Collection.

British Railways No 47998, ex LMS Fowler & Beyer Peacock 2-6-6-2T seen waiting for its turn in the works, circa 1950. For repair the Garratts were allocated to Crewe. Reportedly they were unpopular with the men who had to work on them, or was it simply that they were just different? The introduction of the BR Standard '9F' class hastened the end of the BR Garratts. The 1927 Manchester built No 47998 was withdrawn from Toton (18A) by BR during the week ending 25/08/1956. *Keith Langston Collection*

MOS (WD) 'Austerity' 2-8-0 BR No 90678 is seen ex works but minus its tender in 1965. This locomotive was representative of a class of 533 engines which became BR stock in 1948. Number 90678 was built by Vulcan Foundry Co Ltd in 1944 and withdrawn by BR from Wakefield (56A) during the week ending 24/06/1967. *Rail Photoprints/Brian Robbins.*

The 'Royal Scot' Class 4-6-0 locomotives were originally built at Derby Works and by North British Locomotive Co Ltd. However, when 70 members of the class were rebuilt, also at Derby between 1943 and 1955, the Crewe designed 'Top Feed Tapered LMS 2A' boiler replaced the original parallel type. It was not unusual to see Royal Scot locomotives at Crewe Works pending repair, accordingly as the class were 'not Crewe built' they are represented in this separate section.

Pictured at Crewe whilst being prepared for a trip to North America is a class member presented as LMS No 6100 ROYAL SCOT.
In all probability that locomotive was Derby built class mate No 6152 THE KINGS DRAGOON GUARDSMAN.
Much has been written about the locomotive swop, which by all accounts was never reversed.
Note that the engine has been fitted with a USA style electric headlamp and a smokebox mounted train nameboard. The LMS tender used was non-standard with roller bearing axle boxes.
The Richard Metcalfe Collection.

Rebuilt 'Royal Scot' Class BR No 46100 ROYAL SCOT is seen at Crewe Works after being repainted with British Railways (BR) livery. Note the nameplate commemorating the trip to the 'Century of Progress Exposition', Chicago 1933. This locomotive withdrawn from Nottingham (16A) in October 1962. Norman Preedy Collection.

Rebuilt 'Royal Scot' Class 4-6-0 BR No 46153 THE ROYAL DRAGOON, waits to be reunited with its tender, after an overhaul in January 1961. This locomotive was withdrawn by BR from Annesley (16D) during the week ending 22/12/1962.
Alan H. Bryant/Rail Photoprints Collection.

Rebuilt 'Royal Scot' Class 4-6-0 BR No 46109 ROYAL ENGINEER is seen whilst waiting its turn in the works, circa 1955.
This locomotive was withdrawn by BR from Holbeck (55A) during the week ending 27/12/1962.

Rebuilt 'Royal Scot' Class 4-6-0 BR No 46158 THE LOYAL REGIMENT is awaiting a visit to the paint shop, circa 1960.
This locomotive was withdrawn from Annesley (16D) during the week ending 19/10/1963.

Both images,
Norman Preedy Collection.

Super D 'G2A' class 0-8-0 BR No 49117 is seen outside the works after an overhaul in 1955. This 1906 built locomotive was withdrawn by BR from Swansea Patton Street (87K) in March 1959. Rail Photoprints Collection.www.railphotoprints.co.uk.

Stanier's famous 'Turbomotive' 4-6-2 LMS No 6202 PRINCESS ANNE undergoing repairs at Crewe during May 1939. Records show that the overhaul took 174 days. Upon completion the locomotive was returned to its home depot of Camden (1B) on 07/05/1939. The Pacific was rebuilt as a normal 4-cylinder engine in 1952 but unfortunately was damaged beyond repair in the horrific Harrow railway accident on 8 October 1952. George C. Lander/Rail Photoprints.

Employees.

From a peak of reportedly 20,000 employees the Crewe workforce often varied to reflect the changes in manufacturing techniques and repair workloads, with numbers between 7,000 and 8,000 often quoted as average. Locomotive building in 1848 was recorded as being '1 complete locomotive' per week, but as history has recorded the expansion of the works was ongoing. What is certain is the fact that as the works grew in size and increased its engineering capacity, so did the size of the town. Crewe Locomotive Works was for many years the town's main employer and provider of employment. In 1915, men leaving on war service meant that women were employed in the workshops for the first time.

In 1837 the population of Crewe was quoted as being 1,800 people but only 30 years later in 1871, the population of the town had significantly grown to over 40,000, a large proportion of those residents relied on the railway for their livelihoods. In addition to helping the town with water supply, as mentioned earlier, the works also supplied Crewe with gas. In 1941 the works again required a further increase in gas supply and a substantial extension to the carbonising plant was undertaken. By 1944 over 500 million cubic feet of gas per year was produced with about 30% taken by the railway works, 20% supplied to the nearby Rolls Royce works and approximately 50% to the town.

Shop staff when all work on steam locomotives ended in 1967 was 5,377 employees. The table below shows the breakdown per department.

Shop	Employees	Shop	Employees
Erecting shop	517	Tin shop	38
Unclassified repairs shop	171	Bogie repairs (erecting shop)	107
Diesel test centre	125	Bogie repairs (tender shop)	180
Main machine shop	650	Steel foundry	155
Heavy machine shop	59	Signal shop	48
Power-unit repairs	207	Stores department	140
Traction-motor repairs	132	Millwrights & joiners	329
Auxiliary-motor repairs	105	Boiler house	6
Generator repairs	36	Apprentices & juniors	539
Paint shop	88	Miscellaneous	339
Wheel shop	91	Supervisory (workshops)	332
Plates, fabrication, welding	389	Salaried staff	439
Copper shop	116	Management	39

Oxy-acetylene cutting.

'Knocking Up'. A key is being inserted into the aligned slots cut in the wheel boss and axle. Both images Crewe Works Archive.

Crewe Built Steam Locomotive Classes.
Steam locomotives which were taken into British Railways stock in 1948, or built by British Railways.

LMS Stanier '3MT' 2-6-2T BR 40185 to 40209, 1937/38.
25 engines Crewe built.

BR Number series 40071–40209, total built 139.
Derby Works 40071 – 40184 1935/38.
None preserved.

Power Classification	3P Reclassified 3MT in 1948
Designer Stanier	Company LMS
Driving wheel	5ft 3ins
Boiler pressure	200PSI Superheated
Cylinders	Outside 17½ ins x 26 ins
Tractive Effort	21485lbf
Valve gear	Walschaert (piston valves)

The class were built as an improvement on Fowlers previous underpowered engines of this type. However, these locomotives were also underboiled. Accordingly, 4 of the class were fitted with larger boilers during 1941, and a further 2 in 1956.

Sir William Arthur Stanier (1876 – 1965).
William Stanier was born in Swindon, Wiltshire GB and in 1891 he followed in his father's footsteps and joined the Great Western Railway (GWR). In 1904 he was appointed to the post of 'Assistant to The Divisional Locomotive Superintendent' in London, returning to Swindon in 1912 to become 'Assistant Works Manager' becoming 'Works Manager' in 1920. In 1931 he was famously headhunted by Sir Josiah Stamp, 'Chairman of the London Midland Scottish Railway' (LMS). He served as Chief Mechanical Engineer (CME) of the LMS from I January 1932 until his retirement in 1944. He introduced tapered boilers to the LMS, building on the experience he had gained at Swindon. In all, Stanier gave his name to twelve individual LMS locomotive classes and improved others.
Honours:- Knighted in 1943. Fellow of the Royal Society (F.R.S.) in 1944.

Stanier 3MT 2-6-2T BR No 40186 passes Haymarket with a Corstophine - North Berwick local passenger service, 01/09/1957. This locomotive was withdrawn from Motherwell (66B) in December 1962. David Anderson.

LMS & SDJR Fowler '2P' 4-4-0 BR 40636 to 40660 1931, 40686 to 40700, 1932. 39 engines Crewe built.

BR Number series 40563 – 40700, total built 138.
Derby Works   40563 – 40635 1928/30,
40661 – 40685 1931/32. None preserved.

Power Classification	2P
Designer Fowler	Company LMS/SDJR
Driving wheel	6ft 9ins
Boiler pressure	180PSI Superheated
Cylinders	Inside 19 ins x 26 ins
Tractive Effort	17730lbf
Valve gear	Stephenson (piston valves)

Sir Henry Fowler (1870-1938).

Henry Fowler was born in Evesham, Worcestershire. His railway engineering career began when he commenced an apprenticeship under John Aspinall at the Lancashire & Yorkshire Railway (L&YR) Horwich in 1887. He moved to the Midland Railway (MR) in 1900, in 1905 became Assistant Works Manager, and was promoted to Works Manager two years later. He became Chief Mechanical Engineer (CME) in 1909. From 1915/19 he undertook war work with the RAF becoming 'Assistant Director of Aircraft Production' in 1918. His Midland Railway career is perhaps best remembered for the creation of the iconic 'Royal Scot' class of express passenger locomotives (introduced 1927). After 'Railway Grouping' he became Deputy CME of the newly formed LMS under George Hughes, becoming CME in 1925, retiring in 1933.
Honours:- In 1917 awarded C.B.E. followed by the K.B.E. 1918. Knighted in 1920 for his 'Contributions to the War Effort'.

These 138 successful '2P' 4-4-0 engines were specifically built for the London Midland & Scottish Railway with three additional engines (LMS numbers 575/6 and 580) being built for, and allocated to the Somerset & Dorset Joint Railway (SDJR) who renumbered them in their series 44 - 46.
They were built with reduced boiler mountings giving them a lower loading gauge, thus allowing them to work all over the LMS. Often used in double heading, the 4-4-0s also worked extensively in Scotland on the former lines of the Glasgow & South Western Railway (G&SWR). Withdrawals began in 1959 and by the end of 1962 the class withdrawal was complete.

Unfortunately two of the class LMS numbers 591 and 639 were scrapped in 1934, after being involved in a railway accident at Port Eglinton near Glasgow on 6 September. That head on collision tragically resulted in the death of six passengers and three footplatemen. In addition, a large number of people received injuries some of which were described as serious.

The scene at Port Eglinton 06/09/1934.

LMS S&DJR Fowler Midland design inside cylinder '2P' 4-4-0 BR No 40643 is seen departing Dumfries with a Glasgow & South Western Railway route 4-coach local service for Kilmarnock, in June 1959. This locomotive was withdrawn from Hurlford (30B) on 31/10/1961.
Dave Cobbe Collection/
www.railphotprints.co.uk.

LMS S&DJR Fowler Midland design inside cylinder '2P' 4-4-0 BR No 40649 seen at Eastfield in ex works condition after a visit to Cowlairs Works, circa 1955. This locomotive was withdrawn from Corkerhill (30A) on 18/11/1959. David Anderson.

LMS S&DJR Fowler Midland design inside cylinder '2P' 4-4-0 BR No 40648 at Ayr Motive Power Depot during spring 1953. This locomotive withdrawn from Aberdeen Ferryhill (61B) on 27/09/1961. David Anderson.

LMS S&DJR Fowler Midland design inside cylinder '2P' 4-4-0 BR No 40645 and 'Royal Scot' BR No 46130 THE WEST YORKSHIRE REGIMENT passing Kilmarnock with the 'Thames Clyde Express' on 06/04/1960. Note the ex-Caledonian Railway 'Bow Tie' route indicator on the smokebox door top lamp bracket of the lead engine. The '2P' was withdrawn from Hurlford (30B) on 31/10/1961. Rail Photoprints Collection/A. E. Durrant.

LMS S&DJR Fowler Midland design inside cylinder '2P' 4-4-0 BR No 40646 waits at Luton Bute Street station before taking over the Stephenson Locomotive Society (SLS Midland Area) 'Tour of Seven Branch Lines' charter for the return to Birmingham, on 14/04/1962. This locomotive was withdrawn from Bescot (21B) on 12/05/1962. A. J. B. Dodd/Rail Photoprints.

LMS S&DJR Fowler Midland design inside cylinder '2P' 4-4-0 BR No 40646 seen with SR 'Schools Class' 4-4-0 BR No 30925 CHELTENHAM at Annesley shed on 12/05/1962. The two 4-4-0s worked the RCTS East Midlands Branch 'The East Midlander' on 13/05/1962. The Fowler '2P' was at that time 'officially withdrawn', but was reinstated to work the charter. Ken Gray.

LMS S&DJR Fowler Midland design inside cylinder '2P' 4-4-0 BR No 40696 and S&D 2-8-0 BR No 53805 arrive at Bath Green Park with what is believed to be a Bournemouth West - Huddersfield 'Saturdays Only' service, on 13/08/1960. The '2P' was withdrawn from Bath Green Park S&DJR (then 82F) on 25/05/1962. Rail Photoprints/Hugh Ballantyne.

LMS S&DJR Fowler Midland design inside cylinder '2P' 4-4-0 BR No 40698 departs from Bath Green Park with the 4.35 local service to Templecombe, on 07/09/1957. Note the livestock van in the consist. The '2P' was withdrawn from Bath Green Park S&DJR (then 82F) on 27/07/1960. Rail Photoprints/Hugh Ballantyne.

LMS S&DJR Fowler Midland design inside cylinder '2P' 4-4-0 BR No 40700 is seen between duties alongside the wooden building which was Bath Green Park S&DJR (then 82F) on 27/07/1960. The '2P' was withdrawn from that depot on 28/08/1962. Image courtesy of John Firth.

LMS Ivatt '2MT' 2-6-2T BR 41200 to 41319, 1946/1952. 120 engines Crewe built.

BR Number series 41200 – 41329, total built 130.
Derby Works 41320 – 41329 1952.
Four preserved 41241, 41298, 41312 and 41313.

Power Classification	2P
Reclassified	2MT in 1948
Designer Ivatt	Company LMS
Driving wheel	5ft 0ins
Boiler pressure	200PSI Superheated
Cylinders	Outside 16 ins x 24 ins
Later engines	Outside 16½ ins x 24 ins
Tractive Effort	17410lbf
Later engines	18510lbf
Valve gear	Walschaert (piston valves)

From the 130 locomotives built, only 10 engines appeared before nationalisation (1948). The majority of the class were allocated to depots of the London Midland Region (BR) but a smaller number were allocated to Southern Region depots (BR). Some of the class were originally fitted with taller than normal narrow chimneys, which were considered ugly by some observers, and later engines were fitted with a more traditional wider chimney. In an effort to improve performance, locomotives BR Nos 41290 – 41329 were fitted with slightly larger cylinders. This design later became the basis for the BR Standard 'Class 2' 2-6-2T number series 84000 – 84029. Locomotive BR No 41272 was the 7000th locomotive to be built at Crewe Works, in 1950.

Henry George Ivatt (1886-1976).

H. G. Ivatt was the son of the Great Northern Railway (GNR) engineer Henry Alfred Ivatt and was born in Dublin, Ireland. He started an apprenticeship at LNWR Crewe Works in 1904 and after which, worked in the drawing office later becoming 'Head of Experimental Locomotive Work'. During WW1 he served in France with the 'Office of the Director of Transport'. Returning to England after the war he took up the post of 'Assistant Locomotive Superintendent' with the North Staffordshire Railway (NSR). As part of the 1923 'Amalgamation' the NSR became part of the LMS. From 1932 he served the LMS as 'Divisional Mechanical Engineer Glasgow' until 1937, when he returned to England as 'Principal Assistant for Locomotives' under Stanier. He was appointed 'LMS Chief Mechanical Engineer' (CME) in 1946, a post he held until 1951.

Ivatt 'Class 2' MT 2-6-2T BR No 41202 climbing away from Shepton Mallet towards Cranmore with a Yatton - Witham service, the train has just passed what in modern times is the Mendip Vale terminus of the East Somerset Railway. This locomotive was withdrawn from Edgeley (9B), weekending 26/11/1966. Rail Photoprints/Hugh Ballantyne

Ivatt 'Class 2' MT 2-6-2T BR No 41212 arrives at Stamford Town with the 09.35 from Seaton on 16/09/1965. Stamford Town station was renamed Stamford station in 1966. The mock Tudor buildings are influenced by the nearby Burghley House, designed by Sancton Wood. This locomotive was withdrawn from Leicester Midland (15A) in November 1965.
Rail Photoprints/Hugh Ballantyne.

Having taken water (water column and discarded bag can be seen on the platform) Ivatt 'Class 2' MT 2-6-2T BR No 41214 prepares to leave the delightful station at Torrington, Devon on the former Tarka Valley Railway with the single coach forming the 16.00 train to Halwill Junction, on 05/08/1963. Passenger services from this station ceased in 1965 but milk trains (2 per day) continued until the line was closed completely in 1982. This locomotive was withdrawn from Templecombe S&D (83G) on 02/07/1965. Ian Turnbull/Rail Photoprints.
The Tarka Valley Railway Society was formed in 2008, and in 2023 ran their first train over a short section of track, visit www.tarkavalleyrailway.org

Ivatt 'Class 2' 2MT 2-6-2T locomotives seen on depots and at work.

41217 LONGSIGHT 1959
41220 CHESTER 1960 
41230 LLANDUDNO JCT 1962 
41233 EDGELEY 1965 
41249 EXMOUTH JCT 1964 
41285 CARLISLE 1965 
41277 departing Tutbury for Uttoxeter with single coach train in 1959.

Locomotive withdrawals 41277 in 1962, 41217, 41220, 41233 in 1966 and 41230 in 1967, 41249, 41285 in 1966.
All Images Keith Langston Collection

Push 'n Pull fitted Ivatt 'Class 2' 2MT 2-6-2T BR No 41219 outside the running shed at Northampton (2E), on 30/07/1963. This locomotive was withdrawn from Leicester Midland (15A) during the week ending 09/10/1965. RPC
www.railphotoprints.co.uk

Ivatt 'Class 2' 2MT 2-6-2T BR No 41276 at Llandudno Junction after arriving with the 'Welsh Dragon' a named local service from Rhyl, in the summer of 1959. Note the chimney top damage presumably caused by extra strong emissions from the blast pipe! This locomotive was withdrawn from Barnstaple Junction (72E) during the week ending 31/12/1963. Rail Photoprints /Alan H. Bryant ARPS

Ivatt 'Class 2' 2MT 2-6-2T locomotives on the Somerset & Dorset Joint Railway.

Top. 41291 sits in the centre road at Evercreech Junction, BR Standard Tank 80037 can also seen, 31/07/1965.
Middle. 41296 crossing the GWR mainline at Cole with a short ECS, 11/12/1965.
Bottom. 41307 departing Edington Burtle with a Highbridge-Templecombe service, 11/12/1965.
All 3 locomotives withdrawn from Templecombe (83G) February/March 1966.
All images Hugh Ballantyne/Rail Photoprints.

Ivatt 'Class 2' 2MT 2-6-2T BR No 41245 seen in the Middle Roads at Sheffield Victoria Station circa 1960. This locomotive was withdrawn from Barnstaple Junction (83F) on 28/12/1963. Rail Photoprints Collection.

Ivatt 'Class 2' 2MT 2-6-2T BR No 41291 pauses at Axminster with the LCGB's 'East Devon' Railtour on 07/03/65. This locomotive was withdrawn from Templecombe S&D (83G) during the week ending 20/02/1966. Hugh Ballantyne/Rail Photoprints

Ivatt 'Class 2' 2MT 2-6-2T BR No 41291 is seen again but this time with a commuter service as the 2MT awaits departure from Victoria with the 10.07 service to Tunbridge Wells West, 15/09/1957. Amongst the station adverts can be seen a poster for Michael Todd's film 'Around the World in 80 Days', what's more you could perhaps enjoy watching it with a glass of 'Tizer the Appetizer'. But note that litter was just as much a problem back then, as it is now! Rail Photoprints/Dave Cobbe Collection -C. R. L. Coles.

Ivatt 'Class 2' 2MT 2-6-2T BR No 41287 at Brighton Kemp Town with the RCTS/LCGB 'Sussex Downsman' Railtour on 22/03/1964. This locomotive was withdrawn from Eastleigh (70D) during the week ending 17/07/1966. Hugh Ballantyne/Rail Photoprints.

Ivatt 'Class 2' 2MT 2-6-2T BR No 41299 calls at Baynards with a Guildford - Horsham service, during April 1963. This locomotive was withdrawn from Eastleigh (70D) during the week ending 02/10/1966. Rail Photoprints/John Day Collection.

Ivatt 'Class 2' 2MT 2-6-2T BR No 41319 seen when rostered to 'Station Pilot' duty at London Waterloo on 26/04/1967, just three months before it became surplus to BR requirements. This locomotive was withdrawn from Nine Elms (70A) during the week ending 09/07/1967. Keith Langston Collection.

A Bobby's Job! The signalman at Evercreech North S&DJR prepares to receive the single line token as Ivatt 'Class 2' 2MT 2-6-2T BR No 41307 comes off the branch at Evercreech Junction North Box with the 4pm Highbridge - Evercreech Junction train, 31/07/1965. This locomotive was withdrawn from Templecombe (83G) during the week ending 13/03/1966. Hugh Ballantyne/Rail Photoprints.

Preserved Ivatt 'Class 2' 2MT 2-6-2T BR No 41298 is seen at the Longmoor Military Railway, the occasion being a 1968 public open day at that establishment. The Longmoor Military Railway ceased operations on 31 October 1969. This locomotive was withdrawn from Nine Elms (70A) on 09/07/1967. Keith Langston Collection.

In 2023 the immaculately preserved Ivatt 'Class 2' 2MT 2-6-2T BR No 41298 was based at the Isle of Wight Steam Railway.
This image is courtesy of Peter Skuce.
Visit https: www.iwsteamrailway.co.uk

Preserved Ivatt locomotives at work on the Severn Valley Railway (SVR). 'Class 2' 2MT 2-6-2T BR No 41312 is seen at Waterworks double heading with Ivatt Mogul BR No 46521 as the pair head for Bridgnorth with a train from Kidderminster, in September 1999. Keith Langston.
Visit SVR www.svr.co.uk. The locomotives home base in 2023 was The Mid Hants Railway visit www.watercressline.co.uk.

Crewe Belt System

At the beginning of construction of new engine builds the frames and cylinders were set up on stands at the east end of No 8 shop. That work was carried out under the supervision of a chargehand familiar to that part of the process. Thereafter templates and spirit levels were used, working to the centre lines scribed on the frames. The frame structure was progressively lifted and moved in a westward direction one stage at a time, where at each stage two fitters and an apprentice performed the predetermined operations. Thus the engine under construction moved steadily towards the west end of the shop and when completed it went straight out of the west doors.

'The Belt' in motion. Some workers moved as the plate was exposed, but it is a fascinating image nevertheless. Crewe Works Archive.

Some 30 years later that process developed into what became known as the 'Crewe belt' system, introduced in 1927. For example during locomotive heavy maintenance the working of the Crewe belt system improved efficiency, by controlling the amount of time that a locomotive spent under repair. Each bay of the main shop had two 50-ton, four motor electric overhead cranes which were supplemented by three 10 ton cranes in each of the bays. For maintenance the 'belt' system operated in 10 stages.

> *Over 2 days the locomotive was stripped down to its frames utilising one of the adjacent 'pit' lines.*
- *Stages 1 to 4. The frame ensemble was lifted forward to one of 4 repair stages. It remained in that stage for 4 days where four gangs moved between the stages carrying out their allotted tasks.*
- *Stage 5. On day 7 a repaired bogie and one other wheel pair (or two wheel pairs for a non-bogie engine) were positioned in order that the frame–boiler assembly could be lowered onto them.*
- *Stages 6 to 10. There the locomotives under repair (usually 5 in number) were coupled to each other and a steel hawser, connected to an electrically powered winch sited outside the east end of the shop. The work plan called for activity to cease for 10 minutes after every 7 hours 50 minutes of elapsed work time. The winch then pulled the whole line of engines forward one stage. Thereafter the men at each of the stages resumed work on the next engine. It was this process that gave rise to the name 'belt'.*

The Crewe belt system started with one class of engine, but after a short period it was found that the belt work teams could handle a mix of locomotive classes at the same time. A 'belt' system produced a repaired locomotive every working day, with an engine spending 12 days under repair. After four years of operation the 'full belt' system was found to be too inflexible and so, with the exception of hauling the completed locomotives out of the shop, it was discontinued in favour of movements by crane. However, the Crewe erectors continued to refer to repair/construction lines as 'the belt' for many years.

LMS Hughes/Fowler 5MT 'Crab' 2-6-0 locomotives, under construction on the 'New Engine Belt'. Crewe Works Archive.

LMS Hughes/Fowler 5MT 'Crab' 2-6-0. BR 42730 to 42809 1926/29, 42850 to 42944 1930/1932.
175 engines Crewe built.

BR number series 42700-42944, total built 245.
Three preserved 42700, 42765, 42859.
Horwich Works 42700-42729, 42810-42849 1926/1930.

Power Classification	5F
Reclassified	5MT/6P5F in 1948
Designer Hughes/Fowler	Company LMS
Driving wheel	5ft 6ins
Boiler pressure	180PSI Superheated
Cylinders	Outside 21 ins x 26 ins
Tractive Effort	26580lbf
Valve gear	Walschaert (piston valves)
1931 42818/22/24/25/29 poppet	Lenz rotary cam
1952/4 42818/22/24/25/29	Reidinger poppet

The high running plate gave the 'Moguls' a distinctive look. To stay within the Midland loading gauge the cylinders were placed high and at a sharp angle to the footplate. That feature earned the engines the name 'spiders' which over time was changed to 'crabs'. They were efficient mixed traffic engines which proved to be successful in traffic.

George Hughes (1865 – 1945)
George Hughes was born in Benwick, Cambridgeshire and embarked upon his railway career in 1882 when he became a Premium Apprentice at the LNWR Crewe. He later moved to the Lancashire & Yorkshire Railway (L&YR) where in 1904 he became Chief Mechanical Engineer (CME) and introduced the L&YR locomotive classification system circa 1919. When the L&YR were amalgamated with the LNWR (1922) he became CME of the combined group, and in 1923 progressed to the LMS becoming their first CME, retiring in 1925. He designed the 'Crab' class of locomotives which were built at Horwich in 1926. Hughes had retired by then and those engines were built under the supervision of Fowler.

LMS Hughes/Fowler Horwich 'Crab' 2-6-0 LMS No 2759 on the Hope Valley route in the vicinity of Totley, in 1938. Locomotive BR No 42759 was withdrawn from Gorton (9G) during the week ending 12/01/1963. RPC www.railphotoprints.co.uk.

LMS Hughes/Fowler Horwich 'Crab' 2-6-0 BR No 42736, thundering northbound up Beattock Bank with a heavy train in 1958. Note the narrowness of the Fowler tender in comparison to the width of the locomotives footplate and also the Grangemouth 65F shed plate. This locomotive was withdrawn from Hurlford (67B) on 29/11/1966.
David Anderson.

LMS Hughes/Fowler Horwich 'Crab' 2-6-0 BR No 42742 passes Newton on Ayr with an up freight, on 22/06/1961. This locomotive was withdrawn from Ardrossan (67D) during the week ending 30/07/1962. *Rail Photoprints/J & J Collection - Sid Rickard.*

LMS Hughes/Fowler Horwich 'Crab' 2-6-0 BR No 42748 a Carlisle Kingmoor allocated engine, drifts down the grade from Moorcock tunnel and across Dandry Mire viaduct at Garsdale, Settle & Carlisle route in summer 1958 heading a southbound mixed goods including petroleum tankers. This engine was withdrawn from Gorton (9G) on 10/10/1964.
Gordon Edgar Collection/Rail Photoprints.

LMS Hughes/Fowler Horwich 'Crab' 2-6-0 BR No 42763 approaches Disley Tunnel with a westbound extra, in the summer of 1962. This locomotive was withdrawn from Nottingham (16A) on 06/06/1964. Rail Photoprints/Alan H. Bryant ARPS.

LMS Hughes/Fowler Horwich 'Crab' 2-6-0 BR No 42803 seen northbound on Beattock on 13/04/1955. This locomotive was withdrawn from Motherwell (66B) on 29/11/1966. David Anderson.

LMS Hughes/Fowler Horwich 'Crab' 2-6-0 BR No 42789 in steam in the shed yard at Ayr (67C) in July 1966. This locomotive was withdrawn from Motherwell (66B) on 29/11/1966. Rail Photoprints/Charlie Cross - Gordon Edgar Collection.

Preserved LMS Hughes/Fowler Horwich 'Crab' 2-6-0 BR No 42765 is double heading with a 'GWR 0-6-0T Pannier Tank' during an East Lancashire Railway gala event in 2003. The 'Crab' was withdrawn from Birkenhead Mollington Street (8H) on 10/12/1966. Malcolm Whittaker.
Visit www.eastlancsrailway.org.uk.

**LMS Stanier '5MT' 2-6-0. BR 42945 to 42984
1933/34, 40 engines Crewe built.**

1 locomotive preserved, BR 42968.

The class originally allocated the LMS numbers 13245-13284, renumbered 2945-2984 in 1934.

First of the Finest

Stanier LMS Mogul

Stanier's first mainline steam engine design for the LMS, with tapered boilers. A development of the Horwich 'Crab' 5MT design, but with horizontally mounted smaller diameter cylinders. The engines were originally built with small superheaters, later changed for 21 element units. These engines were often used on heavy excursion trains and also fast freight services. Only 40 were built as they were superseded by the 'Black Five' 4-6-0 class.

Power Classification	5F
Reclassified	5MT/6P5F in 1948
Designer Stanier	Company LMS
Driving wheel	5ft 6ins
Boiler pressure	225PSI Superheated
Cylinders	Outside 18 ins x 28 ins
Tractive Effort	26290lbf
Valve gear	Walschaert (piston valves)

LMS Stanier 2-6-0 '5MT' locomotives almost at the end of their working lives. *Top.* BR No 42945 (first in the class number series) at Leeds City station January 1966, withdrawn 26/03/1966 Heaton Mersey (9F). *Bottom.* BR 42967 at Tiviot Dale in January 1966, also withdrawn from Heaton Mersey (9F) 30/04/1966. Keith Langston Collection.

LMS Stanier 2-6-0 '5MT' BR No 42951 is seen in the locomotive shed at Rugby (2A) during April 1960. This locomotive was withdrawn from Heaton Mersey (9F) on 26/03/1966. Keith Langston Collection.

LMS Stanier 2-6-0 '5MT' BR No 42956 waits to head south from Crewe with a weed killing train, circa 1962. This locomotive withdrawn from Springs Branch Wigan (8F) on 19/09/1964. The pronounced difference in width between the Fowler tender and the locomotives footplate can clearly be seen. Rail Photoprints/Alan H. Bryant ARPS.

LMS Stanier 2-6-0 '5MT' BR No 42965 seen whilst bypassing Birmingham, as it heads a mixed freight along the Sutton Park Line towards Water Orton, on 07/03/1964. This locomotive was withdrawn from Bushbury (21C) on 08/08/1964.
Neville Simms/Ranwell Collection/Rail Photoprints.

Preserved LMS Stanier 2-6-0 '5MT' LMS No 2968 seen on shed at Bridgnorth when returned to steam at the Severn Valley Railway (SVR) in 1990. This locomotive was withdrawn from Springs Branch Wigan (8F) on 31/12/1966. Malcolm Whittaker Collection.

Preserved LMS Stanier 2-6-0 '5MT' LMS No 2968 is seen departing Hampton Loade SVR with a train for Bridgnorth 09/2001. Keith Langston.

Preserved LMS Stanier 2-6-0 '5MT' BR No 42968 is seen departing Bridgnorth SVR with a train for Kidderminster, 10/2011. Keith Langston.

Preserved LMS Stanier 2-6-0 '5MT' BR No 42968 is seen at Kidderminster SVR on 15/10/2011. Keith Langston.

Preserved LMS Stanier 2-6-0 '5MT' BR No 42968 makes a fine sight when seen with a demonstration freight train at Garth-y-Dwr, on the Llangollen Railway (LR), 31.04.2008. Keith Langston.

Visit www.llangollen-railway.co.uk.

Preserved LMS Stanier 2-6-0 '5MT' LMS No 2968 seen hurrying away from a signal check, after passing Rhyl with the Crewe-Holyhead leg of the Past Time Rail 'Ynys Mons Express' 1T60 on 19/10/1997. We can only hope that the occupiers of the nearby dwellings didn't have their lines full of washing! Dave Jones.

LMS & SDJR Fowler '4F' 0-6-0.
**BR 44107 to 44176 1924/1926, 44302 to 44311 1926, 44437 to 44456 1927/28, 44507 to 44556 1928, 44562 to 44576 1937,
165 engines Crewe built.**

BR number series 44027-44606, 580 locomotives.
Two preserved BR 44027 and 44123.
Derby 44027-44056, 44207-44301, 44407-44436, 44577-44606 1924/41.
North British Locomotive Co Ltd 44057-44081, 44382-44406, 44477-44506 1925/27.
Kerr Stuart Co Ltd 44082-44106, 44332-44356 1925/27.
St Rollox Works 44177-44206, 44312-44331, 44467-44476 1924/28.
Andrew Barclay Sons & Co Ltd 44357-44381 1927/28.
Horwich Works 44457-44466 1928.

Power Classification	4F
Designer Fowler	Company LMS/S&DJR
Driving wheel	5ft 3ins
Boiler pressure	175PSI Superheated
Cylinders	Inside 20 ins x 26 ins
Tractive Effort	24555lbf
Valve gear	Stephenson (piston valves)

After 1923 the LMS decided to accept the Midland Railway (MR) 4F '43835 Class' 0-6-0 as a standard freight locomotive. Included in the class total were 5 additional engines built by Armstrong Whitworth specifically for the Somerset & Dorset Joint Railway, BR Nos 44557/61 S&DJR Nos 57 to 61. Officially designated freight engines the class were also regularly used for passenger train workings. In accordance with the wheel arrangement the class earned the nickname 'Duck Sixes'. A total of 580 of the class were taken into BR stock (1948). Significant withdrawals of the class began in 1958 with 11 engines being the last to be retired in 1966.

Above. LMS/S&DJR Fowler '4F' LMS No 4134 is seen shunting at Clifton Down, Bristol in 1936. This locomotive was withdrawn from Coalville (15E) in November 1964. Rail Photoprints Collection.

Right. LMS/S&DJR Fowler '4F' 0-6-0 BR No 44444 is seen shunting at Stockport in 1950. Note, no other BR locomotive had a number with five identical digits. This locomotive was withdrawn from Springs Branch Wigan (8F) during September 1963.

Image courtesy of Ben Brooksbank.

LMS/S&DJR Fowler '4F' 0-6-0 BR No 44156 at Egleton near Rutland has steam to spare, seen with an eastbound freight on the Leicester - Peterborough/Kettering route, on 26/05/1953. This engine was withdrawn from Kettering (15C) during the week ending 22/02/1964. Dave Cobbe Collection/Rail Photoprints.

LMS/S&DJR Fowler '4F' 0-6-0 BR No 44450 seen in ex works condition. The engine was taking a turn as Crewe Works shunter in the summer of 1963, before returning to Mold Junction shed (6B) from where it was withdrawn week ending 08/05/1965. Rail Photoprints - Colin Whitfield.

LMS/S&DJR Fowler '4F' 0-6-0 BR No 44454 takes the freight lines round Derby Midland station with coal empties, during June 1962. This locomotive was withdrawn from Derby (17A) during October 1963. RPC www.railphotoprints.co.uk.

LMS/S&DJR Fowler '4F' 0-6-0 BR No 44512 crossing Braidhurst Viaduct, above the River Calder, with a down freight on 28/03/1963. Locomotive No 44512 was withdrawn from Gorton (9G) during week ending 01/02/1964. J & J Collection - Sid Rickard.

LMS/S&DJR Fowler '4F' 0-6-0 BR No 44523 seen passing Wellow with a local S & DJR route goods bound for Bath Green Park, on 20/09/1958 This locomotive was withdrawn from Bristol Barrow Road (82 E) on 26/07/1963. Rail Photoprints/Hugh Ballantyne.

LMS/S&DJR Fowler '4F' 0-6-0 BR No 44553 runs past the Bristol Barrow Road coaler with an inbound passenger service. The train roofboard reads 'Newcastle - Bristol' which begs the question how far had the 4F worked with the service, was it a Bristol Area failure or possibly a loco change at Gloucester which brought out the '4F', during April 1959. This locomotive was withdrawn from Warrington Dallam (8B) during the week ending 27/10/1962. Rail Photoprints.

LMS Stanier 'Black Five' 5MT 4-6-0.

BR 44658 to 44667 1949, 44718 to 44782 1947/49,
44826 to 44931 1944/46, 44967 to 44981 1946,
45000 to 45019 1935, 45070 to 45074 1935,
45452 to 45471 1938, 241 engines Crewe built.

BR number series 44658-45499, 842 engines built.
18 locomotives preserved, 44767, 44806, 44871, 44901, 44932, 45000, 45025, 45110, 45163, 45212, 45231, 45293, 45305, 45337, 45379, 45407, 45428 and 45491.

Derby Works 44800-44825, 45472-45499 1943/4.
Horwich Works 44668-44717, 44783-44799, 44932-44966, 44982-44999 1945/50.
Sir W G Armstrong Whitworth & Co Ltd 45125-45451 1935/37.
4 engines named, 45154, 45156, 45157, 45158.
Vulcan Foundry Ltd 45020-45069, 45075-45124 1934/5.

LMS  BR

'BLACK STANIERS'

Power Classification	5 MT
Designer Stanier	Company LMS/BR
Driving wheel	6ft 0ins
Boiler pressure	225PSI Superheated
Cylinders	Outside 18½ ins x 28 ins
Tractive Effort	25455lbf
Valve gear	Walschaert (piston valves)
1947,	44767 Outside Stephenson (piston valves)
1948,	44738-44747 Caprotti (poppet valves)
1951,	44686/7 Outside Caprotti (poppet valves)

The Stanier 5MT 'Class 5' 4-6-0 became colloquially known as the 'Black Five', from a combination of the LMS colour and power classification. However, in the early years the 2-cylinder locomotive design was often referred to by the name 'Black Stanier' in order to distinguish it from the designers then LMS red liveried 3-cylinder 'Jubilee' class. The new mixed traffic locomotive specification called for a 2-cylinder 4-6-0 tender engine which could work over at least 70% of the company's routes, whilst being rostered to haul approximately the same amount of freight and passenger trains. All of the boilers for the 415 railway works-built locomotives were manufactured at Crewe, irrespective of which works was responsible for constructing the engines. The new 5MT engines were accurately referred to as Stanier's 'go anywhere locomotives'. They were extremely successful and efficient in traffic, and equally popular with engineers, running shed staff and footplate crews.

During the build period (1934-1951) several modifications to the original design were applied to selected engines. Those changes included visible examples, Caprotti valve gear, high running plates, double chimneys and Outside Stephenson valve gear. Non-visible examples, Steel Fireboxes, Skefco roller bearings and Timken roller bearings on various axles.

LMS Stanier 'Black Five' 5MT 4-6-0 BR No 45006 seen outside the erecting shop at Crewe in 1955, after maintenance and a visit to the paint shop. This locomotive was withdrawn from Crewe South (5B) during the week ending 16/09/1967. Rail Photoprints Collection.

LMS Stanier 'Black Five' 5MT 4-6-0 BR No 44720 is travelling well as it passes Haymarket with an Edinburgh Waverley-Perth express service on 15/10/1955. This is one of 10 locomotives (44718 – 44727) built with a steel firebox. This engine was withdrawn from Dundee Tay Bridge (62B) on 27/10/1966. David Anderson.

LMS Stanier 'Black Five' 5MT 4-6-0 BR No 44751 then allocated to Speke Junction is seen on the depot whilst taking water in the summer of 1962, note the fireman on top of the coal, water valve chain in hand. This is an inside Caprotti variant which was also fitted with Timken roller bearings. This locomotive was withdrawn from Speke Junction (8C) in September 1964. Keith Langston.

LMS Stanier 'Black Five' 5MT 4-6-0 BR No 44755 seen at Crewe Works circa 1949 (**BRITISH RAILWAYS** on tender) fitted with a temporary test/indicating shelter. This is an inside Caprotti variant with double blastpipe and chimney. Note also a steam generator above the cylinder and also the electric light in front of the chimney. This locomotive was withdrawn from Stockport Edgeley (9B) during the week ending 16/11/1963. RPC www.railphotoprints.co.uk.

LMS Stanier 'Black Five' 5MT 4-6-0 BR No 44854 seen departing Dent station (S&C route) with the 3.40pm Bradford - Carlisle service, in August 1964. This locomotive was withdrawn from Normanton (55E) in October 1967.
Dave Cobbe Collection/Rail Photoprints

LMS Stanier 'Black Five' 5MT 4-6-0 BR No 44864 heads west near Dunham Hill (between Frodsham and Helsby) with the Manchester - Llandudno evening commuter service, in April 1959. This locomotive was withdrawn from Edge Hill (8A) during May 1968. Rail Photoprints/R. A. Whitfield.

LMS Stanier 'Black Five' 5MT 4-6-0 BR No 44968 pauses to take water at Crianlarich (upper) station with the 1.20pm working from Mallaig and Fort William to Glasgow Queen Street in the summer of 1960. This locomotive was withdrawn from Motherwell (66B) during May 1964. David Anderson.

LMS Stanier 'Black Five' 5MT 4-6-0 BR No 44922 heads sister engine BR No 45389 (Armstrong Whitworth built) climbing away from Oban towards Glencruitten summit with a service to Glasgow and Edinburgh on 20/09/1957. The Crewe built locomotive was withdrawn from St Rollox (31A) during May 1964. David Anderson.

LMS Stanier 'Black Five' 5MT 4-6-0 BR No 44967 climbs through Glen Falloch with a West Highland Line freight, circa 1957. This locomotive was withdrawn from Dumfries (67E) during May 1964.
Rail Photoprints/John Day Collection.

LMS Stanier 'Black Five' 5MT 4-6-0 LMS No 5071 (BR 45071) is seen passing Elstree with a St. Pancras - Manchester service, during the summer of 1939. This locomotive was withdrawn from Speke Junction (8C) during the week ending 22/07/1967.
Rail Photoprints/C. R. L. Coles -Dave Cobbe Collection.

LMS Stanier 'Black Five' 5MT 4-6-0 BR No 45467 seen in the rock cutting near Cowlairs Junction with a westbound freight, circa 1964. This locomotive was withdrawn from Motherwell (66B) in December 1966. J & J Collection - Sid Rickard.

LMS Stanier 'Black Five' 5MT 4-6-0 BR No 45468 is seen leaving Strathyre with the 7.50am Glasgow to Oban service, BR No 45359 (an Armstrong Whitworth built engine) is seen in the loop with the Stirling to Oban daily goods, on 24/08/1960. The freight will later follow the passenger northbound. The Crewe built locomotive was withdrawn from Grangemouth (31D) during May 1964.
Rail Photoprints/Hugh Ballantyne.

Black Fives at work, LMS Stanier 5MT 4-6-0 BR No 44781 double heading with Vulcan Foundry Ltd built BR No 45046. The pair are seen approaching Dove Holes on 27/04/1968 with 1Z77, a North West Rail Tour charter train. The Crewe built locomotive was withdrawn from Carnforth (10A) during August 1968. Rail Photoprints.

Preserved LMS Stanier 'Black Five' 5MT 4-6-0 BR No 44767 passes Mallaig Junction Yard as it leaves Fort William with the 11.05 Fort William - Mallaig service, on 25/06/1986. This is the unique 'Outside Stephenson Valve Gear' member of the class. Named GEORGE STEPHENSON in preservation (1975). The locomotive was withdrawn from Carlisle Kingmoor (12A) during the week ending 30/12/1967. John Chalcraft/RPC www.railphotoprints.co.uk.

Preserved LMS Stanier 'Black Five' 5MT 4-6-0 LMS No 5000 coupled with 'Standard Tank' BR No 80079 gets away smartly from Chester in April 1980 with a charter train from Hereford to Manchester Victoria. The Crewe built 'Black Five' was withdrawn from Lostock Hall (10D) during October 1967. Keith Langston.

Preserved LMS Stanier 'Black Five' 5MT 4-6-0 BR No 44871 and BR Standard 'Britannia' class Pacific No 70013 OLIVER CROMWELL pass Barton Hill as they leave Bristol with the 10.06 Bristol TM - Preston 'Great Briton III', on 08/04/2010. The 'Black Five' was withdrawn from Carnforth (10A) during the week ending 17/08/1968.
John Chalcraft/RPC www.railphotoprints.co.uk.

LMS Fowler 'Patriot' 5XP, 6P5F, 7P 4-6-0.
BR 45502 to 45519 1932/33, 45523 and 45524 1933,
45529 to 45532 1933, 45536 to 45551 1933/34,
40 engines Crewe built.

Locomotive from which the class name was taken, Derby built.

BR number series 45500-45551, 52 locomotives.
42 engines named.
Derby 45500 and 45501, 45520-45522, 45525-45528, 45533-45535 1930/33.
None preserved.

LMS Baby Scots

Power Classification	5XP reclassified 6P5F in 1951
Rebuilt engines	7P reclassified in 1951.
Designer Fowler	Company LMS
Rebuilt engines Ivatt	Company LMS/BR
Driving wheel	6ft 9ins
Boiler pressure	200PSI Superheated
Rebuilt engines	250PSI Superheated
Cylinders	Three 18"x 26"
Rebuilt engines	Three 17"x 26"
Tractive Effort	26520lbf
Rebuilt engines	29570lbf
Valve gear	Walschaert (piston valves)

In the late 1920s Fowler was working to improve the performance of the 'Claughton' class 4-6-0 locomotives, his 1928 modifications included the fitting of larger boilers. The larger boilers did not solve the problems. Accordingly in 1930 two Claughtons (LMS numbers 5971 and 5902) were completely rebuilt as 3-cylinder engines with three sets of Walschaert valve gear, long travel valves and incorporating the 1928 larger boiler design. The only parts of the earlier engines that remained were the driving wheels.
The new 4-6-0 engines were a complete success and on entering traffic they earned the nickname 'Baby Scots' as they were simillar in design to the original 'Royal Scot' (46100) class, but smaller. Fifty locomotives of the new design were later built by the LMS. The official designation was 'Patriot' class after LMS No 5500 was given that name in February 1937. The first 40 engines were originally considered to be rebuilds of the earlier 'Claughton' class, and carried their names and numbers. Reportedly they did not contain any components from the class that they were supposed to be rebuilds of, being effectivley new engines. In 1934 the whole class was renumbered in the LMS series 5500-5551 (BR 45500-45551), with only 5542-5551 being officially described as new locomotives. A total of 52 'Patriot' 4-6-0 Fowler designed 5XP parallel boiler locomotives were taken into stock by British Railways (BR) in 1948. Under BR, 18 of the class were rebuilt between 1946 and 1949 to the specifications of Crewe Works engineer Ivatt, with larger tapered boilers, double chimneys and new cylinders, those engines were then given a power rating of 7P.

Fowler 'Patriot' 5XP 4-6-0 early LMS No 6000 seen at Kentish Town. Just 10 months old in May 1934. The engine was renumbered to LMS 5538 in 1934 and later to BR 45538, it received the name GIGGLESWICK in 1938. The locomotive was withdrawn from Nuneaton (2B) during week ending 22/09/1962. RPC www.railphotoprints.co.uk.

LMS 'Claughton' class 4-6-0 LMS No 5974 is seen at Kentish Town in 1929.
It was this class of express passenger locomotives which Fowler tried to improve the performance of, and which eventually gave way to the creation of the 'Patriot' class of engines. This locomotive was built in March 1920 and withdrawn in August 1932, as the 'Patriot' class were being constructed.
Rail Photoprints Collection.

Fowler 'Patriot' 5XP 4-6-0 LMS No 5517 (un-named) (BR 45517) arrives at Tamworth with a local service, in 1935. This locomotive was withdrawn from Bank Hall (27A) during week ending 09/06/1962.
Rail Photoprints Collection.

Fowler 'Patriot' 5XP 4-6-0 LMS No 5539 E.C.TRENCH (BR No 45539) on shed at Crewe North (5A) in 1937. This locomotive was withdrawn from Newton Heath (26A) on 16/09/1961. Rail Photoprints Collection.

Fowler 'Patriot' 6P5F 4-6-0 BR No 45504 ROYAL SIGNALS pauses at Pontefract Baghill station with a Birmingham-Newcastle train on 23/05/1959. This locomotive was withdrawn from Warrington Dallam (8B) during March 1962. E. Talbot Collection.

Fowler 'Patriot' 6P5F 4-6-0 BR No 45509 THE DERBYSHIRE YEOMANRY awaits departure from Birmingham New Street with a cross country express, during 1957. This locomotive was withdrawn from Newton Heath (26A) on 12/08/1961.
Rail Photoprints Collection.

Gloucester Eastgate station on 12/08/1959. BR Standard '5' No 73065, LMS 'Patriot' 4-6-0 BR No 45519 LADY GODIVA takes water, and '4F' 0-6-0 BR No 44272 waits to head south on freight. The 'Patriot' was withdrawn from Bristol Barrow Road (82E) on 16/03/1962. J & J Collection - Sid Rickard.

Rebuilt (Crewe Works 1947) 'Patriot Class' 4-6-0 BR No 45529 STEPHENSON (without smoke deflectors) passes Mold Junction with a down express, during 1951. This locomotive was withdrawn from Annesley (16B) on 22/02/1964.
S. D. Wainwright/Rail Photoprints.

Rebuilt (Crewe Works 1946) 'Patriot' 7P 4-6-0 BR No 45530 SIR FRANK REE heads a southbound freight through Berkhamsted circa 1963. This locomotive was withdrawn from Carlisle Upperby (12A) on 01/01/1965. Rail Photoprints Collection.

Rebuilt (Crewe Works 1947) 'Patriot' 7P 4-6-0 BR No 45531 SIR FREDERICK HARRISON seen at Crewe station in July 1964. This locomotive was withdrawn from Carlisle Kingmoor (12A) on 30/10/1965. Alan Fozard.

Official locomotive naming ceremonies would have been quite impressive affairs. The naming of LMS 'Patriot' 5XP LMS No 5504 ROYAL SIGNALS took place at Euston station on 10 April 1937. The locomotive was named by Brigadier Clementi-Smith Colonel Commandant of the regiment, and the LMS was represented by William Stanier. The pair are seen with the engine crew who were wearing their WW1 medals. It was customary to try and roster enginemen for the occasion who had served in the respective regiments. John Magnall Collection.

LMS Stanier 'Jubilee' 5XP, 6P, 7P 4-6-0.
BR 45552-45556 1935, 45607-45654 1934/5,
45665-45742 1935. 131 engines Crewe built.
BR Number series 45552-45742 191 locomotives, all named.
Derby 45655-45664 1934/5.
North British Locomotive Works Co Ltd 45557-45606 1934/5.

Red Stanier

Four locomotives preserved 45593 KOHLAPUR, 45596 BAHAMAS, 45690 LEANDER, 45699 GALATEA.

LMS Stanier 'Jubilee' BR No 45552 SILVER JUBILEE inside Crewe North shed, stored out of use.
Keith Langston Collection.

Power Classification	5XP reclassified 6P5F in 1951
Rebuilt engines	6P reclassified 7P in 1951.
Designer Stanier	Company LMS
Driving wheel	6ft 9ins
Boiler pressure	225PSI Superheated
Rebuilt engines	250PSI Superheated
Cylinders	Three 17"x26"
Tractive Effort	26610lbf
Rebuilt engines	29570lbf
Valve gear	Walschaert (piston valves)

When new in 1935 Stanier 3-cylinder 'Jubilee' class locomotive LMS No 5642 exchanged numbers with LMS No 5552. That locomotive was finished in black enamel with cast chromium plated letters and numerals, chromium plated dome cover and cladding bands. It was named SILVER JUBILEE to celebrate the Silver Jubilee of King George V, thus giving the name to the class.

The last five proposed 4-6-0 locomotives of the Fowler 'Patriot' class (5552-5556) were instead built by Stanier with taper boilers and top feeds thus becoming the first engines of the 'Jubilee' class. The new 5XP locomotives were built concurrently with the 5MT 2-cylinder 'Black Five' class. They were designed to undertake all ordinary mainline work for the LMS with the exception of the heaviest 'top-link' jobs, which Stanier's '8P' locomotives undertook. The original LMS Crimson Lake express passenger livery gave rise to the new engines alternatively being referred to as 'Red Staniers'. When first in traffic the locomotives were tagged as poor performers by the engine crews, often suffering from a shortage of steam. After extensive trials and changes to the blast-pipe and chimney dimensions, their performance was greatly improved. The 'Jubilee' class engines were then heralded a success. Two of the class (BR Nos 45735/36) were rebuilt with larger boilers and double chimneys and classified '7P'. BR took into stock 191 of the class. The first to be scrapped was BR No 45637 WINDWARD ISLANDS which was damaged beyond repair in the 1952 Harrow accident. Final withdrawals began in 1961 and all were out of service before the last year of steam. However, they were the last Stanier designed working express passenger locomotives in BR service.

Preserved LMS Stanier 'Jubilee' 4-6-0 LMS No 5690 LEANDER in LMS livery, seen at Crewe Works during the September 2005 Great Gathering. This locomotive spent all of its working life allocated to Bristol Barrow Road (22A) from where it was withdrawn during the week ending 30/03/1964. Keith Langston.

84 LMS Stanier 'Jubilee' 6P 4-6-0 BR No 45552 SILVER JUBILEE seen on the servicing road at Rugby MPD on 13/07/1963. This locomotive was withdrawn from Crewe North (5A) during the week ending 26/09/1964.
Neville Simms/Ranwell Collection/Rail Photoprints.

LMS Stanier 'Jubilee' 6P 4-6-0 BR No 45618 NEW HEBRIDES leaves Burton on Trent with an up fitted freight on 04/06/1963. This locomotive was withdrawn from Burton on Trent (17B) during the month ending March 1964. Hugh Ballantyne/Rail Photoprints.

LMS Stanier 'Jubilee' 5XP 4-6-0 LMS No 5623 PALESTINE double heads with 'Patriot' LMS No 5545 PLANET. Both Crewe built engines are seen at Gaydon Loops with an up train on 03/04/1937. LMS No 5623 was withdrawn as BR No 45623 from Newton Heath (26A) on 25/07/1964. Edward Talbot Collection.

LMS Stanier 'Jubilee' 6P 4-6-0 BR No 45642 BOSCAWEN approaches Crewe Station with an 'extra' service of the North Wales coast route, the date is 07/08/1958, the first masts in conjunction with the Manchester electrification can be seen. This locomotive was withdrawn from Newton Heath (26A) on 09/01/1965. S. D. Wainwright/Rail Photoprints.

LMS Stanier 'Jubilee' 4-6-0 locomotives from the Norman Preedy Collection.
TOP. BR No 45695 MINOTAUR at Diggle in 1955. Locomotive withdrawn from Farnley Junction (25G) during 02/1964.
MIDDLE. BR No 45701 CONQUEROR at Dillicar Troughs in 1952. Locomotive withdrawn from Newton Heath (26A) 02/1963.
BOTTOM. BR No 45719 GLORIOUS passing Scout Green in 1955. Locomotive withdrawn from Bank Hall (27A) 03/1963.

LMS Stanier 'Jubilee' 6P 4-6-0 BR No 45731 PERSEVERANCE seen near Beattock summit with a down train of vans, spring 1958. This locomotive was withdrawn from Blackpool Central (24E) during the week ending 20/10/1962. David Anderson.

Rebuilt (1942) LMS Stanier 7P 'Jubilee' BR No 45735 COMET, at Willesden MPD 18/09/1961. Note larger tapered boiler, double chimney and smoke deflectors. This locomotive was withdrawn from Annesley (16B) on 03/10/1964.
Ian Turnbull/Rail Photoprints.

Preserved LMS Stanier 'Jubilee' 6P 4-6-0 BR No 45699 GALATEA is seen at Carlisle with 'The Pendle Dalesman' charter on 25/03/2015. This locomotive was withdrawn from Shrewsbury (89A) on 21/11/1964. Michael Halbert.

Preserved LMS Stanier 'Jubilee' 6P 4-6-0 BR No 45690 LEANDER is seen passing Helsby with 1Z50 Liverpool LS- Holyhead and return charter on 16/07/2023. Keith Langston.

LNWR Bowen Cooke 'Claughton' 5P, 5XP 4-6-0.
Introduced 1913 to 1921, 130 engines Crewe built.
62 named

'Claughton Rebuilds' 5XP, 20 rebuilt by Fowler, 1928.
One of the rebuilt class members given a BR number,
LMS 6004 PRINCESS LOUISE (name removed 1935).
Allocated No 46004, but it was never carried.
None preserved.

Power Classification	5P rebuild 5XP
Designer Bowen Cooke	Company LNWR/LMS
Driving wheel	6ft 9ins
Boiler pressure*	175PSI superheated
Fowler rebuilds 1928.	200PSI superheated
Cylinders	Four 15¾" x 26"
Tractive Effort	27070lbf
Fowler Rebuilds 1928	29570lbf
Valve gear	Walschaert (slide valves)
Fowler 1928 rebuilds	Beardmore-Caprotti

1928 Rebuilds - LMS Numbers

Larger Boiler 200PSI s/h	Beardmore-Caprotti
5908	5908
5906	
5910	
5927	5927
5946	5946
5948	5948
5953	
5957	5957
5962	5962
5970	
5972	
5975	5975
5986	
5993	
5999	
6004	
6013	6013
6017	
6023	6023
6029	6029

First of the class No 2222 was named for Sir Gilbert Claughton LNWR Chairman 1911-1921.

The 'Claughton' 4-6-0 locomotives were the last London North Western Railway (LNWR) express passenger design. These undoubtably handsome looking engines, although when introduced were successful, were found to be less so as time in traffic went by, and loadings increased. Because these locomotives were not only poor performers but expensive to maintain, ongoing modifications were tried, but mainly to no avail. It was recorded that the 'Claughton' class engines never bettered the performances of the earlier 'Prince of Wales' and 'George the Fifth' locomotives which they were intended to replace. In 1928 under Fowler 20 of the class received new larger boilers, and of those, 10 were also fitted with Beardmore Caprotti valve gear. This classes association with the later LMS 'Patriot' class engines is well documented. The withdrawal of the class began in 1929, with the last in traffic LMS No 6004 outliving the rest by eight years to be withdrawn as BR 46004 in 1949.

Charles John Bowen Cooke
(1859 – 1920)

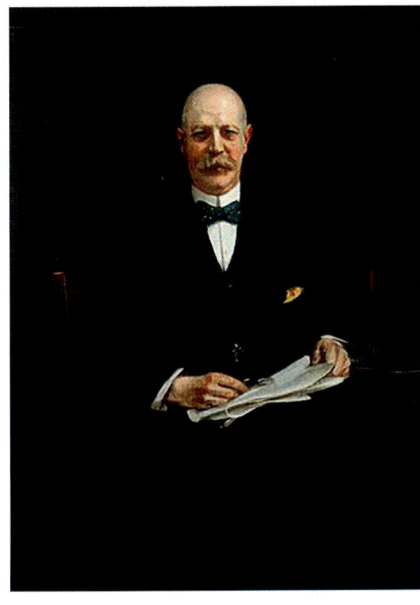

Charles John Bowen Cooke was born in Orton Longueville (then in Huntingdonshire) now Cambridgeshire. He was the son of the Rector of Orton Longueville and he became a premium apprentice at Crewe in 1875, becoming a pupil of Webb in 1878. At the end of his training he became a junior assistant to the Southern Division 'Running Superintendent' and took over that post in 1899. In July 1909 he succeeded George Whale as Chief Mechanical Engineer (CME) of the LNWR. He is famous for his work to improve steam raising efficiency and is credited with the introduction of 'Superheating' and its application to mainline passenger and freight locomotives. He wrote three successive books about locomotive history and development between 1893 and 1902. The eight locomotive classes he introduced were 'George the Fifth' 4-4-0, 'Queen Mary' 4-4-0, 'G' and 'G1' 0-8-0, 'Claughton' 4-6-0, 'Prince of Wales' 4-6-0, 'Prince of Wales 4-6-2T and the '1185' 0-8-2T. He served as Mayor of Crewe 1918-1919 and following some months of ill health died on 18 October 1920. In the 1918 New Years honours he was appointed a Commander of the Order of the British Empire (CBE).
Image from a portrait by James Peter Quinn.

LNWR Bowen Cooke/Fowler 'Claughton Rebuild' 5XP 4-6-0 LMS No 5970 PATIENCE, (large boiler) is seen at Willesden depot during 1928. This locomotive was withdrawn in December 1935. RPC www.railphotoprints.co.uk.

LNWR Bowen Cooke 'Claughton' 5P 4-6-0 LMS 5990 works hard as it passes Camden with a down express, during 1929. This locomotive was withdrawn in April 1935. RPC www.railphotoprints.co.uk.

LNWR Bowen Cooke/Fowler 'Claughton Rebuild' 5XP 4-6-0 LMS No 5975 TALISMAN (large boiler and Beardmore Caprotti valve gear) is seen at London Euston station circa 1930. This locomotive was withdrawn in May 1937. M. Halbert Collection.

LNWR Bowen Cooke/Fowler 'Claughton Rebuild' 5XP 4-6-0 LMS No 5986 (large boiler and smoke deflectors), takes water from Brock Troughs as it heads an up express, circa 1930. This locomotive was withdrawn during November 1935.
Rail Photoprints Collection. www.railphotoprints.co.uk

LNWR Bowen Cooke/Fowler 'Claughton Rebuild' 5XP 4-6-0 LMS No 6004 PRINCESS LOUISE, nameplate removed in June 1935 (large boiler and smoke deflectors) seen at Edge Hill, circa 1948, allocated BR No 46004. This locomotive was withdrawn from Willesden (1A) during April 1949. Rail Photoprints Collection.

LNWR Bowen Cooke/Fowler 'Claughton Rebuild' 5XP 4-6-0 LMS No 6013 (large boiler and Beardmore Caprotti valve gear), seen leaving Crewe with an up service, circa 1931. This locomotive was withdrawn in March 1936.
Rail Photoprints Collection. www.railphotoprints.co.uk

Climbing Beattock
Top. LMS 'Jubilee' 4-6-0 BR No 45677 BEATTY Euston-Perth express during May 1958.
Bottom. LMS 'Jubilee' 4-6-0 BR No 45697 ACHILLES Euston-Perth express during August 1955.
Both images David Anderson.

Climbing Beattock
Top. LMS 'Princess Royal' 4-6-2 BR No 46201 PRINCESS ELIZABETH, Birmingham – Glasgow express summer 1958.
Bottom. LMS 'Princess Royal' 4-6-2 BR No 46209 PRINCESS BEATRICE, Birmingham – Glasgow express during July 1960.
Both Images David Anderson.

LMS Stanier 'Princess Royal' 7P, 8P 4-6-2.
BR 46200 – 46212 1933/35.
13 engines Crewe built, all named.

Two preserved BR 46201 PRINCESS ELIZABETH, 46203 PRINCESS MARGARET ROSE.

LMS Pacific

Power Classification	7P reclassified 8P in 1951
Designer Stanier	Company LMS
Driving wheel	6ft 6ins
Boiler pressure	250PSI Superheated
Cylinders	Four 16¼" x 28"
Turbomotive 1935	None turbine driven (6202)
Rebuilt 1952	Four 16½" x 28" (46202)
Tractive Effort	40285lbf
Rebuilt engine	41540lbf
Valve gear*	Walschaert (piston valves)

*Locomotive 6205 (46205) rebuilt 1947 with 2 sets of Walschaert Gear (piston valves), inside valves operated by rocking shafts. Converted back in 1955.

One of the main tasks facing William Stanier at the LMS was to create a powerful express locomotive for the West Coast Route. His answer was an impressive class of taper boilered 4-cylinder Pacifics. The first two of which appeared in 1933 (LMS 6200/01) were not heralded as the immediate success he thought the design warranted. By evaluating performances gained in traffic with those two engines Stanier amended the design. The original low temperature superheaters were replaced by 32 element units, becoming the standard adopted for the next 10 locomotives to be built (1935). The improvement in performance became immediately obvious as those engines entered traffic. Thus, the design was proved to be successful, and the 'Princess Royal' class performances were thereafter described as being greatly superior to anything previously achieved by an LMS 4-cylinder locomotive. LMS No 6202 was built as an experimental engine (Turbomotive). The customary cylinders and motion was replaced by steam turbines, a large turbine on the left hand side of the locomotive provided forward motion whilst a smaller unit on the right hand side was used for reverse movement. It was the only steam turbine driven locomotive ever to be in regular use in Great Britain. In 1952 a decision was taken to rebuild the locomotive as an ordinary 4-cylinder engine. Unfortunately that engine's working life was cut dramatically short when it was damaged beyond repair in the 8 October 1952 Harrow & Wealdstone railway accident.

Preserved LMS Stanier 'Princess Royal' Pacific LMS No 6201 PRINCESS ELIZABETH passing Pabo Lane, Llandudno Junction with 'Rolex Express' charter 30/06/2019. This locomotive was withdrawn from Carlisle Upperby (12B) 10/1962. Brian Jones.

Turbomotive 'Princess Royal' class Pacific steam turbine driven locomotive BR No 46202 PRINCESS ANNE climbs away from Halton Junction with the 5.25 Liverpool Lime Street - Euston, on 21/05/1949. The late R. A. Whitfield/Rail Photoprints.

LMS Stanier 'Princess Royal' Pacific BR No 46202 PRINCESS ANNE seen at Shrewsbury following a running in turn during August 1952, soon after the rebuild to a normal locomotive, but with slightly larger cylinders. David Anderson Collection.

LMS Stanier 'Princess Royal' Pacific BR No 46203 PRINCESS MARGARET ROSE seen on shed at Polmadie (66A) on 22/04/1956. This locomotive was withdrawn from Carlisle Kingmoor (12A) during the week ending 20/10/1962.
David Anderson.

LMS Stanier 'Princess Royal' Pacific LMS No M6206 PRINCESS MARIE LOUISE (BR 46206) with the 09.30 Glasgow/Edinburgh - Birmingham on Moore Troughs, near Daresbury WCML on 01/08/1948. This locomotive was withdrawn from Camden (1B) during week ending 03/11/1962. Rail Photoprints/the late R. A. Whitfield.

LMS Stanier 'Princess Royal' Pacific LMS BR No 46212 DUCHESS OF KENT seen working hard on Beattock Bank with a Birmingham – Glasgow express on 15/08/1955. This locomotive was withdrawn from Crewe North (5A) on 07/10/1961. David Anderson.

LMS Stanier 'Princess Royal' Pacific BR No 46200 with the RCTS/Stephenson Locomotive Society 'Aberdeen Flyer' railtour on 03/06/1962. No 46200 hauled the Carlisle-Crewe-London Euston leg of this six-locomotive charter.
Top. Locomotive waiting to depart, after a stop at Crewe.
Bottom. The 'Princess Royal' seen reversing the coaching stock out of Euston station. Both images David Anderson.

Preserved LMS Stanier 'Princess Royal' Pacific LMS No 6201 PRINCESS ELIZABETH approaching Penmaenmawr on the North Wales Coast route with a charter for Holyhead on 22/04/2006. Keith Langston.

Preserved LMS Stanier 'Princess Royal' Pacific LMS No 6203 PRINCESS MARGARET ROSE is seen at Crewe North shed (5A) in 1963 after having been cosmetically restored prior to being statically displayed at Butlins Holiday Camp, Pwllheli, North Wales where it remained until 1975, when it moved to the Midland Railway-Butterley. Keith Langston Collection.

Visit www.midlandrailway-butterley.co.uk

Stanier Streamlined Pacifics

LMS Stanier 'Princess Coronation/Duchess' 7P Pacific LMS No 6220 CORONATION in streamlined form at Camden in 1938. The locomotive was painted in 'Caledonian Blue' with horizontal silver bands (aluminium paint) to match the livery of the 'Coronation Scot' coaches, locomotives 6220-6224 (built 1937) carried that livery. This locomotive reportedly held the British steam speed record of 114mph between 1937 and 1938. However, the accuracy of the speed recording was disputed by some railway officials at the time. LMS No 6220 was de-streamlined in November 1946 and withdrawn as BR 46220 from Carlisle Upperby (12B) in April 1963. Rail Photoprints/C. R. L. Coles (Dave Cobbe Collection).

LMS Stanier 'Princess Coronation/Duchess' 7P Pacific LMS No 6229 DUCHESS OF HAMILTON. The streamlined locomotive (in almost new condition) is seen leaving Shrewsbury with a northbound service, during October 1938. LMS locomotives numbered 6225-6229 (built 1938) carried a similar livery style to the first five engines but they were liveried in the companies standard Maroon (Crimson Lake). DUCHESS OF HAMILTON was de-streamlined in December 1947, and as BR No 46229 was withdrawn from Edge Hill (8A) on 15/02/1964. After passing into private ownership it was placed on static display at Butlins Holiday Camp, Minehead, Somerset. The Streamlined Pacific remained there until 1975 when it moved to York, effectively on loan to the National Railway Museum, who eventually purchased the locomotive outright in 1990. In 2009 re-streamlining of LMS No 6229 was completed at Tyseley Locomotive Works, following a Steam Railway Magazine supported public appeal. Rail Photoprints Collection.

LMS Stanier 'Princess Coronation' or 'Duchess' 7P, 8P 4-6-2.
BR 46220-46257 1937/38.
38 engines Crewe built, all named.
Three preserved 46229 DUCHESS OF HAMILTON,
46233 DUCHESS OF SUTHERLAND, 46235 CITY OF BIRMINGHAM.

LMS Stanier 4-6-2

Power Classification	7P reclassified 8P in 1951
Designer Stanier*	Company LMS/BR
Driving wheel	6ft 9ins
Boiler pressure	250PSI Superheated
Cylinders	Four 16½" x 28"
Tractive Effort	40000lbf
Valve gear	Outside Walschaert (piston valves) with rocking shafts

*Locomotives numbers 6256 (LMS) and 46257 (BR) were built in 1947/48 by Ivatt. With roller bearings, altered rear ends and cabs.

In 1937 following on from the success of his 'Princess Royal' class, Stanier introduced an improved design of Pacific locomotives, initially to satisfy the traffic requirements dictated by the LMS London – Glasgow express service (WCML). In order to meet the challenge of Gresley's East Coast Route streamlined 'A4' class the LMS 'Princess Coronation/Duchess' class were built with a streamlined casing. The aesthetic effect was at first a shock to the railway fraternity, many of whom dubbed the engines appearance as being bath-tub like and ugly. However, the new Pacific class were impressive engines which performed well with the heavy West Coast express services. The whole class had been fitted with double chimneys post 1938. In January 1939 LMS No 6229 DUCHESS OF HAMILTON after changing identities with LMS No 6220 CORONATION was sent to the U.S.A., there representing Great Britain at the 'Worlds Fair' in New York. Plans to ship the locomotive back to GB had to be abandoned as the outbreak of WWII intervened causing the locomotive to be stranded in the U.S.A.. The locomotive eventually returned in 1943 and then reverted back to its original identity. The decision to remove the streamlined casings was taken in 1946 when it was decided that the feature added little value at speeds below 90mph. Until boiler changes took place the ex streamlined engines could be identified by a sloping smokebox top, giving rise to the loco spotting term 'semis'. Also after de-streamlining, smoke deflectors were added. From 1960 onwards diesel locomotives began to replace the class on WCML trains, and even though some of the engines were in great condition having only recently been outshopped, BR management nevertheless ruled that the last 20 should be taken out of service by December 1964.

LMS Stanier 'Princess Coronation/Duchess' class 8P Pacific BR No 46220 CORONATION is seen passing the signal box at Balerno Junction with an Edinburgh – Glasgow service on 29/06/1956. It was common practice at the time to roster an ex WCML Stanier Pacific to work these local passenger trains. The sloping top of the smoke box can clearly be seen on this image of a 'semi'. This locomotive was withdrawn from Carlisle Upperby (12B) in April 1963. David Anderson.

LMS Stanier 'Princess Coronation/Duchess' class 8P Pacific BR No 46220 CORONATION is seen on shed at Camden (1B) in April 1961. Note that the locomotive no longer has a sloping smokebox top. Ian Turnbull/Rail Photoprints Collection.

LMS Stanier 'Princess Coronation/Duchess' class 8P Pacific BR No 46221 QUEEN ELIZABETH is seen passing Balerno Junction with a local passenger service circa 1956. Note the Caledonian Railway 'Bow Tie' route indicator carried on the buffer beam middle bracket. This locomotive was withdrawn from Carlisle Upperby (12B) in May 1963. David Anderson.

LMS Stanier 'Princess Coronation/Duchess' class 8P Pacific BR No 46221 QUEEN ELIZABETH is seen whilst being turned at Glasgow Polmadie depot in 1955. David Anderson.

LMS Stanier 'Princess Coronation/Duchess' class 8P Pacific BR No 46223 PRINCESS ALICE is seen passing Balerno Junction with a local passenger service on 26/10/1957. Note that on this occasion the Caledonian Railway 'Bow Tie' is on the top lamp bracket. The smokebox door number plate has a customary Scottish Region blue background. This locomotive was withdrawn from Polmadie (27A) during October 1963. David Anderson.

LMS Stanier 'Princess Coronation/Duchess' class 8P Pacific BR No 46225 DUCHESS OF GLOUCESTER climbing Beattock Bank with a Birmingham – Glasgow express during the late summer of 1959. This locomotive was withdrawn from Carlisle Upperby (12B) during the week ending 12/09/1964. David Anderson.

LMS Stanier 'Princess Coronation/Duchess' class 8P Pacific BR No 46226 DUCHESS OF NORFOLK is seen near Crawford, WCML with a Euston – Perth train circa 1955. This locomotive was withdrawn from Carlisle Kingmoor (12A) during the week ending 12/09/1964. David Anderson.

Streamlined LMS Stanier 'Princess Coronation/Duchess' class 7P Pacific LMS No 6226 DUCHESS OF NORFOLK is seen passing South Kenton with the down 'Mid-Day Scot', during 1938. This locomotive was de-streamlined in June 1947.
D. Cobbe/Collection of the late C.R.L. Coles/Rail Photoprints.

Streamlined LMS Stanier 'Princess Coronation/Duchess' class 7P Pacific LMS No 6227 DUCHESS OF DEVONSHIRE pictured in Crewe Works after having just been passed to enter traffic, in June 1937. The livery was Crimson Lake and Gold and it was the third engine to be painted in that style. This locomotive was de-streamlined during August 1946 and was withdrawn as BR No 46227 from Polmadie (27A) on 29/12/1962. Mike Bentley Collection.

LMS Stanier 'Princess Coronation/Duchess' class 8P Pacific BR No 46231 DUCHESS OF ATHOLL seen above at Slateford Junction circa 1955. In the image below BR No 46231 is seen passing Elvanfoot at speed during May 1960. This locomotive was withdrawn from Polmadie (27A) during the weekending 29/12/1962. Both Images David Anderson.

LMS Stanier 'Princess Coronation/Duchess' class 8P Pacific BR No 46235 CITY OF BIRMINGHAM waits with a down express adjacent to Rugby No 4 signal box circa 1959. This locomotive was withdrawn from Crewe North (5A) during the week ending 12/09/1964. A preserved locomotive, static exhibit 'Thinktank' Birmingham Science Museum. N.E. Stead/Keith Langston Collection.

LMS Stanier 'Princess Coronation/Duchess' class 7P Pacific BR No M6236 CITY OF BRADFORD (became BR 46236) is seen at Crewe North MPD on 10/07/1948. Note **BRITISH RAILWAYS** on the tender and the 'M' prefix. For a short period BR used prefixes to indicate the regional origin of the locomotives it inherited i.e. M – Midland Region. This locomotive was withdrawn from Carlisle Kingmoor (12A) during the week ending 14/03/1964. RPC www.railphotoprints.co.uk.

The Coronation Scot 1937 – 1939

Introduced in 1937 the streamlined Coronation Scot train was a 'total concept' designed and built by the LMSR (London Midland & Scottish Railway) in order to achieve fast travel between London Euston and Glasgow Central. The specially designed streamlined Stanier Pacific locomotives were built at Crewe Works and the carriage sets at the LMS Wolverton Works. The locomotives were LMS numbers 6220, 6221, 6222, 6223 and 6224.

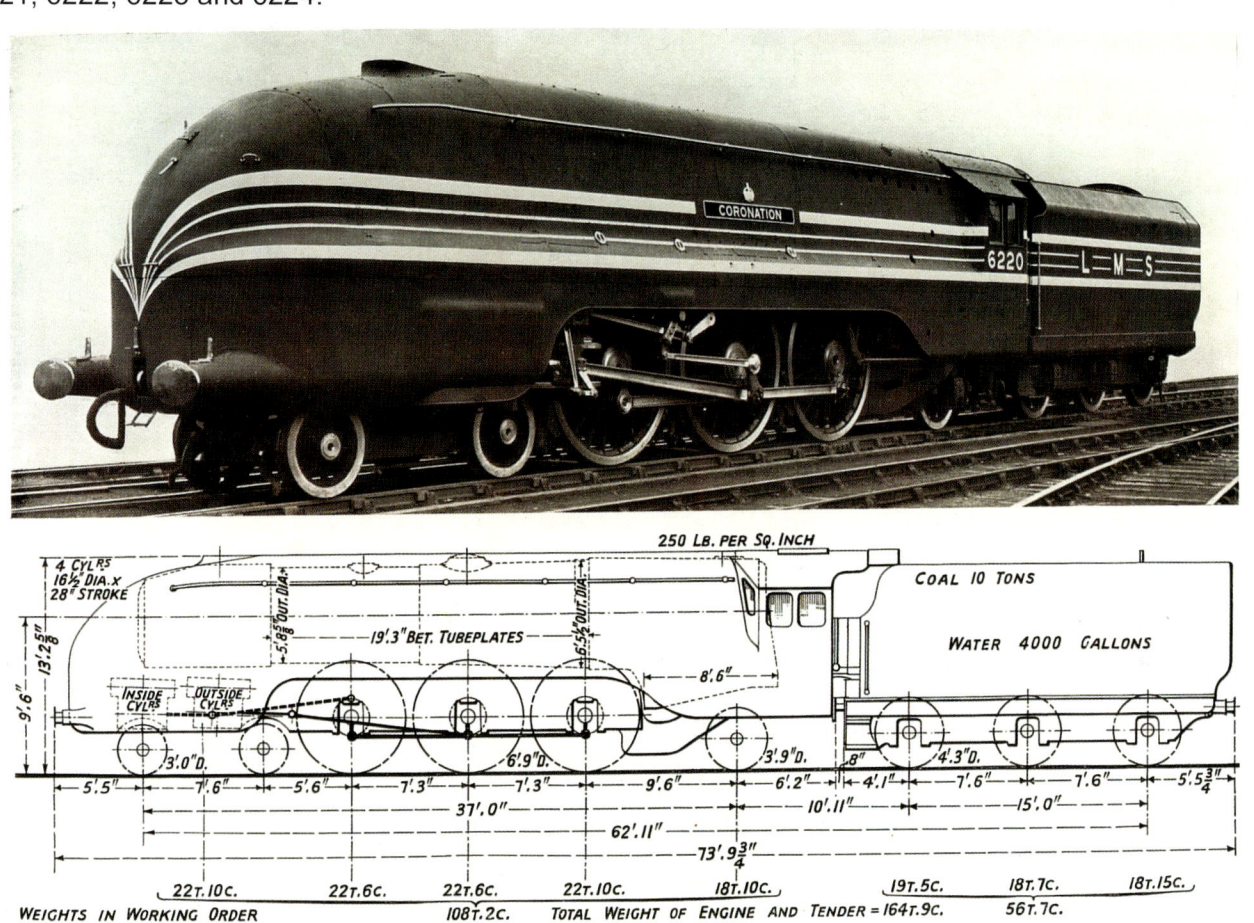

Image and drawing Richard Metcalfe Collection.

In a 1937 article for the railway press the LMS gave the following summation of the new Coronation Scot concept. *'New 4-6-2 locomotives and nine-coach trains for working the LMSR accelerated 6½ hour express service between Euston and Glasgow to commence on July 5th. The engines are streamlined, and, like the train sets are finished throughout in blue and silver'.*

The name 'Coronation Scot' was chosen in order to commemorate the Coronation of King George VI (12 May 1937). The stylised named train ran from July 1937 until the start of World War II in 1939.
There were three nine coach train sets which all comprised of an 18 seat Corridor First Brake, a 22 seat Corridor First, a 42 seat Vestibule First Class Diner, a Kitchen Car, 42 seat Vestibule 3rd, a 42 seat Vestibule 3rd Diner, a Kitchen Car, 42 seat Vestibule 3rd (Diner) and a Corridor Third Brake.
The coach interiors were designed in what for the age, was a very modern style and lighting was plentiful and imaginatively installed. The coach sets were finished in different types of wood and each in their own way, described by some observers as futuristic.
The first 'Coronation Scot' train was presented to the press at Crewe Station following a lunch at the nearby Crewe Arms Hotel on Tuesday 26 May 1937, six weeks before the first planned mainline run. The press were said to have received the new rail travel concept well and they were very complimentary about it. The press party were welcomed to the launch by none other than Mr W. A. Stanier, and the after lunch inspection of the new train was conducted under the guidance of Mr F. A. Lemon, who was at that time Crewe Works Superintendent.
The actual running time planned for the new service was 385 minutes and that timing included a 5-minute stop at Carlisle. That schedule called for an average speed of 62.6 mph to be achieved, a speed that meant both the locomotives and its crews must perform at, or at least near, the top limit of their capabilities.

A reproduction of one of the several LMS advertising posters from the time of the Coronation Scot launch. The service only operated on weekdays and during the summer months. Keith Langston Collection.

Anglo - Scottish named trains, hauled by Stanier Princess Coronation/Duchess class locomotives.

London Euston – Glasgow Central via the West Coast Main Line.
First titled run 11 July 1927, title withdrawn 9 September 1939.
Reinstated 16 February 1948.
Last titled run 1 June 2002.
BR type 6 headboard with either black, red, maroon or light blue background.
In the early 1960s steam haulage gave way to diesel, which in 1966 began to give way to AC Electric, firstly London Crewe (then Diesel north) but eventually electric throughout.
For many years the daily train departed from both termini at 10.00.

LMS Stanier 'Princess Coronation/Duchess' class 8P Pacific BR No 46232 DUCHESS OF MONTROSE seen passing Carstairs with a 'down' Royal Scot in the summer of 1955. This locomotive was withdrawn from Polmadie (27A) in December 1962. David Anderson.

LMS Stanier 'Princess Coronation/Duchess' class 8P Pacific BR No 46242 CITY OF GLASGOW passing Elvanfoot with a 'down' Royal Scot during July 1959. This locomotive was withdrawn from Polmadie (66A) on 18/10/1963. David Anderson.

Anglo-Scottish named trains, hauled by Stanier Princess Coronation/Duchess class locomotives.

London Euston – Glasgow Central via the West Coast Main Line.
First titled run 26 September 1927, title withdrawn 9 September 1939.
Reinstated 26 September 1949.
Last titled run 13 June 1965.
One of 8 aluminium headboards which included a hyphenated name.
Headboard style choice included either a red or light blue background.

LMS Stanier 'Princess Coronation/Duchess' class 8P Pacific BR No 46225 DUCHESS OF GLOUCESTER at Carstairs Junction, with an 'up' Mid-Day Scot in 1953. The locomotive was at that time in BR Blue livery. David Anderson.

London Euston – Glasgow Central via the West Coast Main Line.
First titled run 17 June 1957.
Non stop run, the successor to the Coronation Scot.
Last titled run 4 September 1964.
Headboard style shown is a later version with narrower shields.

LMS Stanier 'Princess Coronation/Duchess' class 8P Pacific BR No 46239 CITY OF CHESTER seen near Weaver Junction with the down 'Caledonian', 24/07/1958. This locomotive was withdrawn from Crewe North (5A) during the week ending 12/09/1964. Rail Photoprints/R. A. Whitfield.

LMS Stanier 'Princess Coronation/Duchess' class 8P Pacific BR No 46245 CITY OF LONDON seen at Crewe station in August 1963. This locomotive was withdrawn from Crewe North (5A) during the week ending 12/09/1964. R.A. Whitfield/Rail Photoprints.

LMS Stanier 'Princess Coronation/Duchess' class 8P Pacific BR No 46246 CITY OF MANCHESTER is seen topping Beattock Summit with a southbound Mid-Day Scot, the locomotive was in BR Blue livery on 04/07/1959. This locomotive was withdrawn from Camden (1B) during the week ending 26/01/1963. David Anderson.

LMS Stanier 'Princess Coronation/Duchess' class 8P Pacific BR No 46248 CITY OF LEEDS (not named until 1941) prepares to head south from Crewe, circa 1960, presumably after the gentleman has finished coupling up!. This locomotive was withdrawn from Crewe North (5A) during the week ending 05/09/1964. Rail Photoprints/Alan H. Bryant ARPS.

LMS Stanier 'Princess Coronation/Duchess' class 8P Pacific BR No 46251 CITY OF NOTTINGHAM is seen at Crewe North MPD in the summer of 1963. Note the overhead catenaries, the electricity had arrived! This locomotive was withdrawn from Crewe North (5A) during the week ending 12/09/1964. Keith Langston Collection.

LMS Stanier 'Princess Coronation/Duchess' class 8P Pacific BR No 46253 CITY OF ST. ALBANS is seen on Christelton water troughs departing Chester for Euston with a Bowater Group staff special on 03/06/1959. This locomotive was withdrawn from Crewe North (5A) during the week ending 26/01/1963. R.A. Whitfield/Rail Photoprints.

LMS Stanier 'Princess Coronation/Duchess' class 7P Pacific BR No 46257 CITY OF SALFORD, a BR engine as the last of the class to be built, is seen at Weaver Junction with an up express during 1949. Note the electric lights and **BRITISH RAILWAYS** on the tender. This locomotive was withdrawn from Carlisle Kingmoor (12A) during September 1964. R.A. Whitfield/Rail Photoprints.

Preserved LMS Stanier 'Princess Coronation/Duchess' class 8P Pacific BR No 46229 DUCHESS OF HAMILTON (prior to being re-streamlined). The Duchess is pictured soon after passing Rhyl with a North Wales Coast Express charter during the late summer of 2009. Dave Jones.

Preserved LMS Stanier 'Princess Coronation/Duchess' class '8P' Pacific LMS No 6233 DUCHESS OF SUTHERLAND ex Holyhead pauses at Llandudno Junction with a charter in the summer of 2009. Brian Jones.

LMS Ivatt '2MT' 2-6-0, 46400 to 46464 1946/50.
65 engines Crewe built.
Number series 46400 – 46527, 128 built.
Darlington Works 46465 – 46502 1951/52.
Swindon Works 46503 – 46527 1952/53.
7 preserved 46428, 46441, 46443, 46447 and 46464 (Crewe) 46512 and 46521 (Swindon).

IVATT MOGUL

Power Classification	2F reclassified 2MT in 1948
Designer Ivatt	Company LMS/BR
Driving wheel	5ft 0ins
Boiler pressure	200PSI Superheated
Cylinders	Outside 16" x 24"
Darlington/Swindon built	Outside 16½"x24"
Tractive Effort	17410lbf
BR built engines	18510lbf
Valve gear	Walschaert (piston valves)

Just two years before 'Nationalisation' H.G. Ivatt introduced two new designs of lightweight locomotives for use on cross country and branch line workings. The tender engine version was the 2-6-0 (Mogul) 4MT '46400' class (the sister 2MT tank locomotive type being the '41200' class). Examples of both classes first appeared from Crewe Works during December 1946. Of the 128 tender engines eventually built, only 20 appeared as LMS locomotives in the companies number series 6400 – 6419. Those examples had a smaller lower pitched boiler, and lower running plate than the BR built engines. The successful Ivatt design was later adopted as a British Railways standard locomotive type appearing as the BR Standard 2MT '78000' class in 1952. Most of the 2-6-0 Ivatts, which earned the nickname 'Mickey Mouse', were based on the London Midland Region of BR, but some were also allocated to depots on the Western, Eastern, North Eastern and Scottish Regions. There were several variations in chimney design, 46400 – 46464 carried short LMS type, 46465 – 46489 carried tall tapered types, 46490 – 46527 carried BR tall parallel types. One engine, BR No 46424 was fitted with an experimental 'stove-pipe', chimney in 1951, earning it the nickname 'The Spout'.

LMS Ivatt '2MT' 2-6-0 BR No 46401 is seen in ex works condition at Wigan, presumably after the Gloucester Barnwood allocated engine had completed a 'running in' turn from Crewe Works, circa 1959. This locomotive was withdrawn from Buxton (9D) during May 1966. RPC www.railphotoprints.co.uk.

LMS Ivatt '2MT' 2-6-0 BR No 46403 leaves Oakham with a westbound local service, on 26/05/1953. This locomotive was withdrawn from Stirling (65J) on 29/06/1964. Dave Cobbe/Rail Photoprints.

'Shift change at Manchester Victoria station'. LMS Ivatt '2MT' 2-6-0 BR No 46411 awaits its next turn of duty, circa 1961. This locomotive was withdrawn from Newton Heath (26A) in January 1967. Keith Langston Collection.

LMS Ivatt '2MT' 2-6-0 BR No 46440 and footplate crew seen at Derby in May 1964. This locomotive was withdrawn from Northwich (8E) in March 1967. Rail Photoprints Collection. Inset, at Northwich May 1967. Keith Langston.

LMS Ivatt '2MT' 2-6-0 BR No 46443 with Stephenson Locomotive Society (SLS) Derby – Sheffield at Lenton South Junction on 03/07/1954. This locomotive was withdrawn from Newton Heath (9D) during March 1967. *Michael Halbert Collection.*

LMS Ivatt '2MT' 2-6-0 BR No 46451 is seen light engine between turns, at Kilmarnock station on 19/05/1965. This locomotive was withdrawn from Hurlford (67B) during December 1966. *Rail Photoprints Collection.*

In 1952 BR No 46460 was transferred to work on the St. Combs Light Railway in Scotland where it was fitted with a cow catcher. The engine is seen at Fraserburgh in the company of LNER 'D40' BR No 62276 ANDREW BAIN on 17/05/1953. The Mogul was withdrawn from Ayr (67C) on 24/08/1966. RPC www.railphotoprints.co.uk.

LMS Ivatt '2MT' 2-6-0 BR No 46462 is seen light engine between turns, at St. Margarets in 1959. This locomotive was withdrawn from Bathgate (64F) during August 1966. Rail Photoprints Collection.

Preserved LMS Ivatt '2MT' 2-6-0 BR No 46443 is seen crossing the River Mawddach (ex Barmouth) with the Cardigan Bay Express but carrying 'The Red Dragon' headboard on 31/08/1987. Keith Langston.

LNWR Webb '1P' 2-4-2T 1890 to 1897.
43* engines Crewe built.
* Crewe built 180 small wheeled versions followed by 160 of the '1P' larger wheeled engines.

BR number series 46601 – 46757.
None preserved.

Power Classification	1P
Designer Webb	Company LNWR
Driving wheel	5ft 8½ ins
Boiler pressure	150PSI
Cylinders	Inside 17" x 24"
Tractive Effort	12910lbf
Valve gear	Allan straight link (slide valves)

Webb produced two classes of 2-4-2Ts for the London and North Western Railway which were developed from an earlier '58092' class of tank engines. The first withdrawals from the 160 larger wheeled versions began in 1921 and only 43 engines became BR stock. The design of the locomotives remained virtually unchanged throughout their working lives (apart from renumbering and painting) they even retained their vintage style LNWR chimneys. However, 20 of the class were motor fitted for working 'push pull' trains. Scrapping continued at a steady rate, with none of the class surviving beyond the end of 1954.

Francis William Webb (1836 – 1906).

Francis Webb was born in Tixhall Rectory, near Stafford and he was the second son of William Webb, Rector of Tixall. As a young man Webb showed an aptitude for all things mechanical. On 11 August 1851, aged 15 he joined Crewe Works as an articled pupil of Francis Trevithick. After his period of training he joined the drawing office staff, becoming Chief Draughtsman on 1 March 1859. In 1861 Webb became Works Manager and Chief Assistant to John Ramsbottom. As Works Manager he was responsible for the start of steel production at Crewe. In order to gain experience in steel making techniques Webb moved to the Bolton Iron & Steel Company in 1866. Returning to Crewe at the invitation of Richard Moon (LNWR Chairman) he became Locomotive Superintendent in 1870 (title changed to Chief Mechanical Engineer 1871). He was responsible for 29 locomotive designs and the production of 2,366 engines, earning him the title 'King of Crewe'. Webb was also responsible for the remodelling of Crewe station including the freight train underpasses. He was an Alderman on both Crewe Town Council and Cheshire County Council and was twice elected as Crewe Mayor, also serving as a magistrate. Francis William Webb is remembered as a major benefactor of the Webb Orphanage, and supporting the council in the creation of Queen's Park. After a serious illness Webb retired in May 1903. The 2005 Great Gathering at Crewe Works was organised in the name of 'Webb Crewe Works Charity Fund'.

LNWR Webb '1P' 2-4-2T BR No 46701 built in 1894 demonstrates it still has a hi-speed capability, as it propels two auto-coaches south over Moore Troughs WCML, circa 1950. This locomotive was withdrawn from Bangor (6H) during February 1953.
Rail Photoprints/R. A. Whitfield.

LNWR Webb '2P' 0-6-2T 1898 to 1902.
80* engines Crewe built.
*15 of the class to BR stock.
BR Numbers 46876, 46878, 46881, 46883, 46899, 46900, 46906, 46909, 46912, 46917, 46920, 46922, 46924, 46926 and 46931.
None preserved.

Power Classification	2P
Designer Webb	Company LNWR
Driving wheel	5ft 2½ ins
Boiler pressure	150PSI
Cylinders	Inside 18" x 24"
Tractive Effort	15865lbf
Valve gear	Joy (slide valves)

These engines were a passenger version of Webb '58880' class Coal Tanks. Built with larger wheels they were known as Watford Tanks or 18" Passenger Tanks, specifically designed for suburban passenger work. Withdrawals began in 1920 with none surviving past 1953.

Webb 0-6-2T 'Watford Tank' BR No 46912 is seen adjacent to Northampton No 2 signal box on 11/6/1949. This locomotive was withdrawn from Monument Lane (3E) in December 1951.
H.C. Casserley/LNWR Society.

LNWR Webb Bissel Truck '1F' 0-4-2PT 1896 to 1901.
20* engines Crewe built.
*2 of the class to BR stock.

BR numbers 47862 and 47865.
None preserved.

Power Classification	1F
Designer Webb	Company LNWR
Driving wheel	4ft 5½ ins
Boiler pressure	150PSI
Cylinders	Inside 17" x 24"
Tractive Effort	16530lbf
Valve gear	Stephenson (slide valves)

A Bissel truck (Bissel bogie or pony truck) is a single-axle bogie which pivots towards the centre of a steam locomotive and enables it to negotiate curves more easily. The system was first developed in 1857 by the American engineer Levi Bissell. With that type of bogie the single axle is able to turn about its vertical axis and swing radially, an advantage when the locomotive had to negotiate tight track curves in limited spaces. Withdrawals of the class commenced in 1929 with only two locomotives passing into BR stock as Crewe works shunters.

LNWR Webb Bissel Truck '1F' 0-4-2PT BR No 47862 at Crewe Works on 17/05/1951. This locomotive was withdrawn during November 1956. RPC www.railphotoprints.co.uk

LNWR Bowen Cooke '6F' 0-8-2T 1911 to 1917.
30* engines Crewe built.
* Only 9 of the class to BR stock.
47875, 47877, 47881, 47884/5, 47887/8, 47892, 47896.
None preserved.

Power Classification	6F
Designer Bowen Cooke	Company LNWR
Driving wheel	4ft 5½ ins
Boiler pressure	170PSI
Cylinders	Inside 20½" x 24"
Tractive Effort	27240lbf
Valve gear	Joy (slide valves)

Bowen Cooke designed this '6F' locomotive as a side tank version of his 'G1' class 0-8-0 freight locomotives. Basically the type was intended as shunting engines for large Marshalling Yards etc but on frequent occasions they were rostered for mainline duties. The design incorporated the saturated type 'Precursor' class boiler with lagged ends, round topped fireboxes and sloping coal bunkers. The main wheels were coupled by overlapping rods with the third pair of wheels being flangeless to allow tighter turns to be made. The lever operated 'Joy' valve gear was fitted instead of the more normal Ramsbottom system. The slim style Cooke buffers were later replaced with a standard Webb pattern type. Braking was provided by steam but an additional vacuum system allowed the engines to work with fitted freight or passenger stock. The first of the class was withdrawn in 1934.

LNWR Bowen Cooke '6F' 0-8-2T with LNWR No 736 is seen at Willesden depot during July 1921. E. Talbot Collection.

LNWR Bowen Cooke '6F' 0-8-2T LNWR No 1592 is seen in the depot at Willesden during March 1922 and still in lined livery. The '6F' 0-8-2T is one of three classes of LNWR engines which displayed the companies initials on the side tanks, as normal practice. E. Talbot Collection.

LNWR Beames '7F' 0-8-4T 1923 to 1924.
30 engines Crewe built.
Only 14 of the class to BR stock,
Allocated numbers 47930/33, 47936/39, 47948, 47951, 47954, 47956 and 47958/59. Only two engines 47931 and 47937 received their BR numbers. None preserved.

Power Classification	7F
Designer Beames	Company LNWR/LMS
Driving wheel	4ft 5½ ins
Boiler pressure	185PSI superheated
Cylinders	Inside 20½" x 24"
Tractive Effort	29645lbf
Valve gear	Joy (piston valves)

Many of these Beames built engines did not appear until after Grouping in 1920. They were very similar in design to the Bowen Cooke '6F' 0-8-2 locomotives but with the addition of an extra pair of trailing wheels to accommodate a larger bunker. The aforementioned Bowen Cooke locomotives were essentially intended for yard shunting whilst the more powerful Beames engines were built specifically to work over the steeply graded Abergavenny – Merthyr line in South Wales. These engines were often referred to as the 'tank' version of the 7F 'G2' class.

Hewitt Pearson Montague Beames (1875 – 1948)

Hewitt Pearson Montague Beames was born in Monkstown, Dublin, Republic of Ireland. Educated at Corrig School, Kingstown, County Dublin, (now Dún Laoghaire), firstly at Dover College, and afterwards at Crawley's Military Academy. His railway career started at Crewe Works (LNWR) when he became a premium apprentice in 1895, later taken on as a pupil by Francis Webb in 1898. He then became a junior assistant to the Works Manager, but in January 1900 he left in order to serve with 'Paget's Horse' in the Boer War. After 18 months in South Africa he returned to Crewe, and in 1910 became Chief Personal Assistant to Cooke. However in October 1914 (WWI) he again left, this time to join the Royal Engineers. He was recalled to Crewe in April 1916 becoming 'Works Manager and Personal Assistant' to Cooke. Later serving as 'Assistant Chief Mechanical Engineer', and in November 1920 he became 'Chief Mechanical Engineer' (CME). This was a short lived appointment as upon the amalgamation of the LNWR with the Lancashire & Yorkshire Railway (LYR) George Hughes was appointed CME. Beames became LNWR 'Divisional Mechanical Engineer, Western Division and after grouping, 'Mechanical Engineer Crewe'. In December 1930 he was made 'Deputy Chief Mechanical Engineer' under Ernest Lemon. Beames retired on 30 September 1946. He was awarded the CBE by King George VI in 1946.

LNWR Beames '7F' 0-8-4T LMS No 1189 is seen on depot whilst between turns in South Wales circa 1925. This was the third of the class to be built. The LMS commenced withdrawing this class in 1944. E.Talbot Collection.

LNWR Beames '7F' 0-8-4T LMS No 7949, built at Crewe circa September 1923. Note the 8A shed plate. This engine did not survive to be allocated a number in the BR system. The locomotive is seen in 1947 at Edge Hill where it is in the company of 'Jubilee' 4-6-0 BRITISH COLUMBIA. Michael Halbert Collection.

LMS/War Department (WD) Stanier '8F' 2-8-0.
48000 to 48012, 48016 to 48018, 48020, 48024, 48026 1935/37, 48096 to 48175 1938/43, 48301 to 48330 1943/44. 48775 1937.
130 engines Crewe built.

BR Number series 48000 – 48775, with gaps, 852 built.

7 preserved 48151, 48173, 48305, 48431, 48518*, 48624, 48773.

*48518 became the donor for new build locomotives GWR 1014 and LMS 5551.

North British Locomotive Co. Ltd 48176-48225 1942, 48246-48285 1940/42, 48773/74 1940.
Vulcan Foundry Co. Ltd 48027-48029, 48033, 48035-48037, 48039, 48045/46, 48050, 48053-48057, 48060-48065, 48067, 48069/70, 48073-48085, 48088-48090, 48092-48095 1936/37.
Beyer Peacock Co. Ltd 48286-48297 1940/42.
Horwich Works 48331-48399, 48490-48495 1944/45.
Swindon Works 48400-48479 1943/45.
Darlington Works 48500-48509 1948/1950, 48540-48559 1945, 48730–48752 1945/46.
Doncaster Works 48510-48539 1944/45, 48753-48772 1945/46.
Eastleigh Works 48600-48609 1943, 48650-48662 1943/44.
Ashford Works 48610-48612, 48618-48624, 48671-48674 1943/44.
Brighton Works 48613-48617, 48625-48649, 48663-48670, 48675-48729 1943/44.

Stanier 8F 'Engine of War'

Power Classification	8F
Designer Stanier	Company LMS/WD/
Driving wheel	4ft 8½ ins
Boiler pressure	225PSI Superheated
Cylinders	Outside 18½" x 28"
Tractive Effort	32440lbf
Valve gear	Walschaert (piston valves)
Engines 48000–48011 originally built with domeless boilers.	

The LMS fleet of ageing freight engines were proving totally inadequate for the volume of work which they faced in the 1930s. This led the company to urgently seek a design of modern heavy goods locomotive which would solve all their problems. Their CME Mr Stanier came up with the solution. The first of his new '8F' class 2-8-0 locomotives were literally delivered 'straight off the drawing board', and became an instant success. Stanier had taken the experience gained in the construction, and subsequent performances of his highly successful 'Black Five' class, and applied those criteria, in a scaled up form to his new freight locomotive design. The '8F' engines in addition to fulfilling all that was required of them on freight duties, were found to also ride well at speed and were therefore often used on passenger trains, running at speeds of up to 60mph. They had a spacious cab and their reliability ensured a regional acceptability by both engine crews and depot managers.

LMS WD Stanier '8F' 2-8-0 BR No 48119 seen on the former Cheshire Lines route with a freight ex Hawarden in the summer of 1965. This locomotive was withdrawn from Edge Hill (8A) during December 1967. Keith Langston.

The advent of WWII saw the class pressed into intensive service not only at home but also overseas, and thus the Stanier '8F' became Britain's 'Engine of War'. The wartime government ordered that a further 208 of the engines be built, and shipped the whole batch overseas. Several were lost at sea aboard torpedoed ships but the majority made it to the Middle East, and after being converted to oil burners they served on vital supply routes, even ranging as far afield as the Soviet Union. At the time of railway nationalisation in 1948, all the British based 8F's and also 39 engines which had returned from overseas duty, were taken into stock by the BR London Midland Region. Many '8F' engines stayed abroad and worked in Egypt, Iran, Iraq, Israel, Italy, Turkey and the Lebanon, with some locomotives working into the 1980's. In 1957 British Railways held a stock of 666 of the type and in the main they continued to serve the company well, right to the end of the steam era. Locomotive BR No 48616 became the first to be scrapped following a derailment in 1960 and withdrawals 'proper' began with BR No 48009 leading the way to the scrapyards in 1962. Approximately 150 examples survived into the last year of BR steam and the last two retired were BR Nos 48318 and 48773, from Rose Grove depot (10F) in Lancashire, on 4 August 1968.

Left.
A cab view of LMS/WD Stanier '8F' 2-8-0 BR No 48309. Bath Green Park Driver Harold Binford at the helm of the Stanier '8F' between Chilcompton Tunnel and Masbury. The engine was being driven flat out because of vacuum pressure differences between the locomotive and the train brakes, which meant that the brakes were not fully released, 04/04/1965. This locomotive was withdrawn from Bath Green Park (82F) during March 1966.
Hugh Ballantyne/Rail Photoprints.

Below.
LMS/WD Stanier '8F' 2-8-0 BR No 48018 on shed at Northampton (1H) circa 1964. This locomotive was withdrawn from Crewe South (5B) during October 1967. Keith Langston Collection.

LMS/WD Stanier '8F' 2-8-0 BR No 48005 seen approaching Disley Tunnel with a rake of tractors, for export out of Liverpool Docks, 1962. This locomotive was withdrawn from Heaton Mersey (9F) during the week ending 19/03/1966.
Rail Photoprints/Alan H. Bryant ARPS.

LMS/WD Stanier '8F' 2-8-0 BR No 48011 is seen at Foxhall Junction, Didcot with a train of covered wagons, possibly carrying steel/automotive products on 03/04/1965. This locomotive was withdrawn from Patricroft (9H) during the week ending 27/05/1967. David Anderson.

LMS/WD Stanier '8F' 2-8-0 BR No 48121 is departing Woodford Halse with a southbound Great Central route coal working, 12/10/1963. This locomotive was withdrawn from Sutton Oak-Peasley Cross (8G) during the week ending 22/04/1967. Neville Simms/Ranwell Collection/Rail Photoprints.

LMS/WD Stanier '8F' 2-8-0 BR No 48162 is seen back at its 'birthplace' during a 1963 visit to the paint shop. This locomotive was withdrawn from Royston (55D) on 24/06/1967. RPC www.railphotoprints.co.uk.

LMS/WD Stanier '8F' 2-8-0 BR No 48309 seen departing Bath Green Park station with the 1.10pm service to Templecombe on 09/10/1965.This locomotive was withdrawn from Bath Green Park (82F) during March 1966. Hugh Ballantyne/Rail Photoprints

131

LMS/WD Stanier '8F' 2-8-0 BR No 48309 pictured at Shepton Mallet (S&DJR) with the Locomotive Club of Great Britain (LCGB) 'The Wessex Downsman Railtour' on 04/04/1965. There were six locomotives used on this charter which started and ended at London Waterloo. This locomotive hauled the Bath Green Park to Bournemouth West leg. Hugh Ballantyne/Rail Photoprints.

LMS/WD Stanier '8F' 2-8-0 BR No 48307 is seen approaching Winwick Junction with a short engineers train on 28/08/1965. This locomotive was withdrawn from Patricroft (9H) during the week ending 02/03/1968. Hugh Ballantyne/Rail Photoprints Collection.

LMS/WD Stanier '8F' 2-8-0 BR No 48319 with sister engine BR No 48652 (Eastleigh built) and Black Five BR No 45260 (Armstrong Whitworth built) stand together at Bolton shed, all awaiting their journey to the scrapyard, 25/05/1968. The Crewe built locomotive was withdrawn from Bolton (9K) during June 1968. RPC www.railphotoprints.co.uk.

LMS/WD Stanier '8F' 2-8-0 BR No 48326 with empty ICI Northwich limestone hoppers, pictured climbing away from New Mills in lightly falling snow, winter 1960. This locomotive was withdrawn from Sutton Oak-Peasley Cross (8G) during July 1966.
Rail Photoprints/ Alan H. Bryant ARPS.

LMS/WD Stanier '8F' 2-8-0 BR No 48329 'clanking' through Tiviot Dale with a train of empty 'windcutters'. The '8F' looking and sounding rough, and the falling rain did not improve this 1967 vision. This locomotive was withdrawn from Heaton Mersey (9F) on 04/06/1968. Keith Langston.

Preserved LMS/WD Stanier '8F' 2-8-0 BR No 48305 is seen at the Churnet Valley Railway in 2004. This locomotive was withdrawn from Speke Junction (8C) during the week ending 13/01/1968. David Gibson.

Visit www.churnetvalleyrailway.co.uk .

LNWR Whale '19in Goods' 4-6-0 1906 to 1909.
170 engines Crewe built.
3 to BR stock (1948).

BR number series 48801, 48824, 48834.
These number were allocated but not applied. None preserved.

Whale 'Experiment Goods'

Power Classification	4F
Designer Whale	Company LNWR
Driving wheel	5ft 2½ ins
Boiler pressure	175PSI
Cylinders	Outside19" x 26"
Tractive Effort	22340lbf
Valve gear	Joy (piston valves)

LNWR Whale '19in' Goods LMS No 8780 at Crewe South shed on 15/09/1935. www.DevaBob.com

George Whale's 4-6-0 freight locomotive was basically the equivalent of the LNWR 'Experiment' class 4-6-0 of 1905. The company built 170 of these engines at Crewe Works between 1906 and 1909. The first withdrawals of the class began in 1931 and only three locomotives came into BR stock. They never received their allocated BR numbers and were all withdrawn by the end of 1949. Boiler changes took place during their working lives and records show that at least one locomotive was fitted with a Belpaire firebox type boiler.

George Whale (1842-1910)

George Whale was born in Bocking, Essex. His railway career began at LNWR Wolverton Works under James Edward McConnell in 1858. When, in 1862 the LNWR made the decision to concentrate locomotive building at their Crewe site under John Ramsbottom, they transferred approximately 400 workers to Crewe and Whale was included in that number. At Crewe he progressed firstly to become a member of the drawing office staff and in 1867 he moved to join the locomotive running staff, and in 1898 Whale was then made responsible for the overall running of all LNWR locomotives. Upon the retirement of Francis Webb he took over the role of 'Locomotive Superintendent' which he held until his retirement in 1908. George Whale was responsible for the introduction of 5 locomotive classes, 'Precursor', 'Experiment' Precursor Tank' '19in Goods' and also the 'G' class 0-8-0 engines which were built after his retirement.

LNWR Whale '19in Goods' 4-6-0 LMS No 8834 is pictured at Willesden in 1933. This locomotive was allocated the BR No 48834 which it never carried. Note the Belpaire firebox style boiler. This engine was withdrawn from Crewe in December 1948 and scrapped in the following January. RPC www.railphotoprints.co.uk

LNWR 'SUPER D'

LNWR 'G1' 6F & 'G2A' 7F, 0-8-0, 1892 to 1918.
To BR 1948 'G1' 123 engines, 'G2A' 319 engines, all Crewe built.
BR Nos 48892 to 49394 with gaps.
48892-48899, 48901-48915,
48917/18; 48920-48922, 48924-48927,
48929-48936, 48939-48945, 48948,
48950-48954, 48962, 48964, 48966,
49002-49117, 49119-49181, 49183-49205,
49207-49214, 49216-49335, 49337-49373,
49375-49379, 49381-49394.
None preserved.

'G1' Power Classification '	6F
Designers	Webb/Whale/Bowen Cooke/Beames
Company	LNWR
Driving wheel	4ft 5½ ins
Boiler pressure	160PSI superheated
Cylinders	Inside 20½" x 24"
Tractive Effort	25640lbf
Valve gear	Joy (piston valves)

'G2A' Power Classification '	7F
Designers	Webb/Whale/Bowen Cooke/Beames
Company	LNWR
Driving wheel	4ft 5½ ins
Boiler pressure	175PSI superheated
Cylinders	Inside 20½" x 24"
Tractive Effort	28045lbf
Valve gear	Joy (piston valves)

Firstly the name 'Super D' was used to cover all three main variants of the 0-8-0 freight locomotive classes of which the LNWR and then the LMS built a total of 572 engines (see separate listing of 'G2' class engines). During the period of the 0-8-0 type builds it was customary for locomen to generally refer to large boilered freight engines as being a D. Later when superheated boilers were used that colloquial name changed to 'Super D'.

Superheating. Steam created in the boiler at a specified pressure and temperature is known as saturated steam, as it is in contact with the water. By applying additional heat that steam can be raised to a higher temperature (by using one of the many variations of steam engine superheating technology). Basically the principle involves taking the saturated steam and then passing it through a series of heater elements situated in the boilers large flue tubes. The resultant steam when fed to the cylinders is then hotter, drier and as a result has greater expansive qualities, thus producing more power.

The 'Super D' earned the reputation of being a very noisy locomotive and many railwaymen claimed that they could be heard miles away with their 'two loud, then two gentle' exhaust beats with the second of the loud beats being noticeably louder than the first. In addition, the distinctive sound made by the type of 'Joy' valve gear used on these engines in conjunction with the constant ringing of the side rods reportedly made them audibly unique amongst goods engines. The combined classes gave great service and although withdrawal from traffic started in 1947 several engines earned a reprieve and were in fact given heavy overhauls due to a shortage of freight engines in the period immediately after WWII. All of the 'G1' locomotives had been withdrawn prior to 1955 with the 'G2A' engines being in traffic longer and the last three being withdrawn in 1963/4.

LNWR 'G1' 6F 0-8-0 No 1217. This locomotive was originally built as an 'A' class with a round topped firebox. It is seen here post 1924, then with a Belpaire boiler. The 'Super D' class engines never carried BR style smoke box number plates.
E. Talbot Collection.

Webb, Whale, Bowen Cooke and Beames in turn carried out new builds, redesigns and modifications to the original 0-8-0 freight engines, between 1893-1936. These included building/converting 0-8-0 types to 3 and 4-cylinder compound locomotives, which were later all rebuilt as either 2-cylinder simple types or superheated types. Both round topped and Belpaire boilers were used. In 1893 Webb introduced a prototype 0-8-0 simple engine, which became BR 49011, withdrawn in 1949. The Whale/Bowen Cooke 'G' class came into being between 1910 and 1912, with 160PSI non-superheated boilers. They were followed between 1912 and 1918 with the Bowen Cooke 6F 'G1' class engines. Beames introduced the 7F 'G2' class locomotives with 175PSI boilers in 1921/22. Many of the 'G1' class engines were then rebuilt with higher pressure boilers (175PSI) from 1936 onwards becoming the 7F 'G2A' class.

G1 0-8-0 'Super D' locomotives.
Top. LNWR No 2551 built 'A' class 1896, 'D' class 1907, 'G1' class 1925. Withdrawn as BR No 49034, Patricroft (10C) 1962.
Middle. LMS No 9275 built 'G1' class 1917 seen in 1936, 'G2A' class 1939. Withdrawn as BR No 49275 Bescot (3A) 1961.
Bottom. LMS No 9199 built 'G1' class 1912 seen in 1939, 'G2A' class 1947. Withdrawn as BR No 49199 Patricroft (10C) 1962.
Images. No 2551 E.Talbot Collection. No 9275 Rail Photoprints Collection. No 9199 George C. Lander/Rail Photoprints.

LNWR 'G1' 6F 0-8-0 LMS No 9196 seen with an up goods at Wreay on 23/12/1938. This locomotive became a 'G2A' in 1947 and was withdrawn from Carnforth (24L) during October 1961. E.E. Smith/Michael Halbert Collection.

LNWR 'G1' 6F 0-8-0 LMS No 9354 seen at Northampton on 17/07/1938. This locomotive became a 'G2A' in 1946 and was withdrawn from Ryecroft Walsall (3C) during March 1956. E.E. Smith/Michael Halbert Collection.

LNWR 'G2A' 7F 0-8-0 BR No 49046 passes Ashford Bowdler (near Ludlow) with a southbound freight, during the summer of 1951. This locomotive was withdrawn from Hereford (85C) during March 1957. Rail Photoprints/C. R. L. Coles (Dave Cobbe Collection).

LNWR 'G2A' 7F 0-8-0 BR No 49094 seen with the afternoon goods between Diggle and Stockport on 14/07/1951. This locomotive was withdrawn from Bletchley (1E) during December 1962. A. Bendell/Michael Halbert Collection.

LNWR 'G2A' 7F 0-8-0 BR No 49234 seen after taking on water and coal at Patricroft depot during May 1959. This locomotive was withdrawn from Stafford (5C) during June 1960. RPC www.railphotoprints.co.uk

LNWR 'G2A' 7F 0-8-0 BR No 48895 seen in its last months of service whilst shunting wagons of scrap at Wolverhampton, on 11/05/1964. Note, even 16 years after Nationalisation that the tender still carries 'LMS', just visible under the grime.
This locomotive was withdrawn from Bescot (2F) during December 1964. Basil Roberts/Richard Pelham Collection/Rail Photoprints.

LNWR 'G2' 7F 0-8-0 1921 to 1922.
60 engines Crewe built.

BR numbers 49395-49454.
1 preserved BR No 49395.

'G2' Power Classification	7F
Designer	Beames
Company	LNWR
Driving wheel	4ft 5½ ins
Boiler pressure	175PSI superheated
Cylinders	Inside 20½" x 24"
Tractive Effort	28045lbf
Valve gear	Joy (piston valves)

LNWR Super D
Beames 'G2'

The Beames 'G2' 0-8-0 locomotives were a development from the 'G1' class and the inclusion of a superheated 175PSI boiler elevated them to power class 7F. Many of these engines were fitted with Belpaire boilers in place of round topped designs. The 0-8-0 types were the main heavy freight engines of the LNWR and were the only types to become BR stock in any great numbers. Withdrawals began in 1959 and only 3 of the class remained in service in 1963 and none beyond that year.

The preserved locomotive, BR number 49395 is listed as part of the National Collection (NRM) and it was withdrawn from Buxton (9D) in November 1959. The most recent refurbishment of this historically important freight locomotive was undertaken by LNWR Heritage at Crewe. That work was funded by the companies then principle Pete Waterman O.B.E.. The 'G2' returned to steam in 2005 and has since that date visited several preserved railways.

LNWR 'G2' 0-8-0 BR No 49431 is seen coaled up and ready for traffic at Rugby depot circa 1953. This locomotive was withdrawn from Springs Branch Wigan (8F) during December 1962. Ben Brooksbank.

LNWR 'G2' 0-8-0 LMS No 373. Ross pop safety valves were first fitted in 1924. Otherwise this round top firebox locomotive is in almost original condition. This locomotive was withdrawn as BR No 49407 from Bescot (2F) during November 1964.
E. Talbot Collection.

Preserved LNWR Super D 'G2' 0-8-0 BR NO 49395 seen at LNWR Heritage Crewe shortly after being restored to working order in the summer of 2005. Keith Langston.

Preserved LNWR Super D 'G2' 0-8-0 BR NO 49395 arrives at Consall Forge, Churnet Valley Railway with a train for Cheddleton during 2006. David Gibson.

LMS Fowler '7F' 0-8-0 1929 to 1932.
175 engines Crewe built.
BR numbers 49500-49674.
None preserved.

Power Classification	7F
Designer	Fowler
Company	LMS
Driving wheel	4ft 8½ ins
Boiler pressure	200PSI superheated
Cylinders	Inside 19½" x 26"
Tractive Effort	29745lbf
Valve gear	Walschaert (piston valves)

Fowler '7F' aka Baby Austin

This class of 0-8-0 freight locomotives were introduced by Fowler as a development of the LNWR 'G2' class. The engines were both powerful and economical, however the bearing surfaces were found to be undersized and the engines frequently 'ran hot' also causing problems with the motion. As a consequence many were withdrawn from service before the LNWR engines which they were built to replace. The type earned the nicknames Baby Austin or Austin 7 after the popular car produced at the same time.

Fowler 7F 0-8-0 LMS No 9630 with a lengthy mixed freight in West Yorkshire during 1948. This locomotive was withdrawn from Wakefield (25A) during March 1949. www.railphotoprints.co.uk.

Fowler '7F' 0-8-0 BR No 49592 on depot at Wigan L&Y (27D) during 1952. This locomotive was withdrawn from Newton Heath (26A) on 16/05/1959. Rail Photoprints Collection.

LNWR Bowen Cooke 'Prince of Wales' 4P 4-6-0, 1911 to 1922.

4 engines allocated BR numbers, 58000 QUEEN OF THE BELGIANS, 58001 LUSITANIA and 58002 - 58003. 135 engines Crewe built.
None preserved.

Power Classification	4P
Designer	Bowen Cooke
Company	LNWR
Driving wheel	6ft 3ins
Boiler pressure	180PSI superheated
Cylinders	Inside 20½" x 26"
Tractive Effort	22290lbf
Valve gear	Joy (piston valves)

The Bowen Cooke 'Prince of Wales' class 4-6-0 locomotives totalled 246, of which 135 were built at Crewe, 91 were built by Wm. Beardmore & Co. and 20 by North British Locomotive Co. They were used as general purpose mixed traffic locomotives and regularly worked heavy mainline West Coast services. The six coupled engines were a superheated version of the 'George the Fifth' 4-4-0 class. Originally built with round topped fireboxes, many were later rebuilt with Belpaire fireboxes. The first of the class were withdrawn in 1933 and of the six which came into BR stock (LMS Nos 25648, 25673, 25722, 25752, 25787 and 38827) only four were allocated BR numbers.

Top. LNWR 'Prince of Wales' 4P 4-6-0 LMS No 5624 THOMAS CAMPBELL seen in 1928, withdrawn in February 1937.
Bottom. LNWR 'Prince of Wales' 4P 4-6-0 LMS No 5662 ANZAC seen in 1930, withdrawn in August 1936.
Both images RPC www.railphotoprints.co.uk.

LNWR Webb 'Special Tank' 2F 0-6-0ST 1870 - 1880.
Designed Ramsbottom built Webb.
258 engines Crewe built.
243 to LMS stock LMS Nos 7220 to 7457.
5 locomotives to BR stock. CD3, CD6/7/8 and 3323.
None preserved.

Power Classification	2F
Designer/Builder	Ramsbottom/Webb
Company	LNWR
Driving wheel	4ft 5½ ins
Boiler pressure	140PSI
Cylinders	Inside 17" x 24"
Tractive Effort	15430lbf
Valve gear	Stephenson (slide valves)

The 'Special Tanks' were designed especially for shunting work by the LNWR. The first was withdrawn in 1920 and the ones which came into BR stock were all withdrawn before 1959. The engines with a CD prefix shunted at Wolverton, Crewe and Earlstown carriage works but No 3323 retained its original LNWR number and worked as a general shunting engine at Crewe.

John Ramsbottom (1814-1897) was born in Todmorden. He first joined the LNWR in 1846 and in 1862 he became Chief Mechanical Engineer, (CME). Shortly after, Crewe installed their first Bessemer steel works. Ramsbottom developed the 'split metal piston ring', and all reciprocating engines use them to the present day. Retiring in 1871 he became a consultant/director for the Lancashire & Yorkshire Railway (LYR).

LNWR Webb 'Special Tank' 0-6-0ST No 3323 fitted with cab roof and circular smokebox door. Seen standing beside the former Ramsbottom drawing office, Crewe Old Works yard on 16/07/1933. LNWR Society.

LNWR Whale 'Rail Motor' 0-4-0T 1905 to 1906.
7 Rail Motors built, steam engines at Crewe and coach bodies at Wolverton Works.
7 steam railcars to LMS stock Nos 10694 to 10700.
1 Rail Motor to BR stock renumbered 29988.
None preserved.

Power Classification	None
Designer	Whale
Company	LNWR
Driving wheel	3ft 9ins
Boiler pressure	175PSI
Cylinders	Inside 9½" x 15"
Tractive Effort	4475lbf
Valve gear	Stephenson (slide valves)

As the 1900s began railway companies started to examine ways of reducing costs on low volume traffic branch lines, Accordingly in 1905 the LNWR brought into traffic their own version of a 'Rail Motor', with the steam power unit being incorporated in the coach. These units could be driven from either end and therefore the driver could communicate with his fireman via a bell system. In 1933 No 10697 was renumbered 29988 outlasting the rest of the class by 15 years. The unit was officially withdrawn in November 1948 but it had not worked since February 1947 following damage as the result of a collision.

LNWR Whale 'Rail Motor' 0-4-0T No 29988 is seen in these two views at Moffat station, Dumfries & Galloway, then the terminus of the short branch line from Beattock. LNWR Society.

LNWR Webb 18" Goods 'Cauliflower' 2F 0-6-0 1880 to 1902.
310 engines Crewe built. LMS Nos 8315 to 8624.
75 engines to BR stock, 58362 to 58430.
6 withdrawn before numbers allocated.
None preserved.

Power Classification	2F
Designer	Webb
Company	LNWR
Driving wheel	5ft 2½ ins
Boiler pressure	150PSI
Cylinders	Inside 18" x 24"
Tractive Effort	15865lbf
Valve gear	Joy (slide valves)

This 2F goods class were colloquially known as 'Cauliflowers' because originally they carried the LNWR Crest on the centre splasher and some considered that looked like a cauliflower. The class were amongst the most famous British 0-6-0 locomotive types. They were designed for fast goods and as such had larger diameter wheels than the 'Coal Engines'. It was not unusual to see the class rostered to local passenger train workings. Built with round topped boilers many were later rebuilt with Belpaire firebox type boilers. The first was withdrawn in 1922 with 7 engines surviving into 1954.

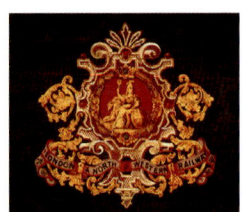

LNWR Webb 18" 2F 0-6-0 goods (Cauliflower) BR No 58396 is seen at Workington depot circa 1950.This locomotive was withdrawn from Widnes (8D) in August 1953.
Ben Brooksbank Collection.

LNWR Webb 'Saddle Tank' (Box Tank) 2F 0-6-0ST 1873 to 1892. 45 engines Crewe built (rebuilt by Whale 1905-1907 from former 2F 'Coal Engine' class.)
LMS Nos 7458-7502. 1 engine allocated BR No 58870.
None preserved.

Power Classification	2F
Designer	Webb
Company	LNWR
Driving wheel	4ft 5½ ins
Boiler pressure	150PSI
Cylinders	Inside 17" x 24"
Tractive Effort	16530lbf
Valve gear	Joy (slide valves)

The Saddle Tanks were created by Whale from mainline life expired 0-6-0 tender engines (Webb 17"Coal Engines). Also known as 'Box Tanks', they were a development of Webb 'Special Tanks' primarily intended for use as shunting engines. The locomotive which came into BR stock was retired in December 1948, only ever having worked at Crewe.

Above. Webb Saddle tank LNWR No 2079 became LMS 7460 and was withdrawn in November 1930.

Left. The official view of Webb Saddle tank LNWR No 808 taken at Crewe on 23 June 1905. This engine became LMS 7475 and was withdrawn March1938.
Both images KL Collection.

LNWR Webb '17" Coal engine LNWR No 3209 is seen passing through Manchester London Road station circa 1915 with freight. This locomotive was transferred to the War Department (ROD) in 1917 (WW1) and did not return to Britain. It is believed to have last worked in Palestine where it was probably broken up circa 1921. LNWR Society.

A delightful shot of Webb '2F' 0-6-0 LMS No 28091 (BR 58321) as it sits under a water tower at Crewe Works on 07/05/1939, The locomotive was withdrawn from service during August 1953 having earnt its keep for some 80 years and 6 months.
George C. Lander/Rail Photoprints

LNWR Webb 'Chopper Tanks' 1P 2-4-0T 1876 to 1885.
50 engines Crewe built. 15 to LMS 6420 to 6434.
1 engine allocated BR No 58092.
None preserved.

Power Classification	1P
Designer	Webb
Company	LNWR
Driving wheel	4ft 8½ ins
Boiler pressure	150PSI
Cylinders	Inside 17" x 20"
Tractive Effort	13045lbf
Valve gear	Allan straight link (slide valves)

Webb 2-4-0T 1P 'Chopper Tank' LNWR No 2238 is seen at Buxton circa 1915. A.G. Ellis Collection/E.Talbot.

This '2234' class of Webb 2-4-0T engines were built at Crewe Works from 1876 and led to a later LNWR Webb design, the 2-4-2T referred to as the '4ft 6in' passenger tank locomotive. The whole class was withdrawn by 1936 however, locomotive LMS No 6428 was taken into BR stock and allocated the number 58092. It outlasted all the others by many years and worked on the Cromford & High Peak Railway until it was withdrawn in March 1952.

LNWR Webb 'Coal Engine' 2F 0-6-0 1873 to 1892.
500 engines Crewe built. 227 to LMS 8088 to 8314.
46 engines to BR stock, 58320-58361, 4 withdrawn before numbers allocated.
None preserved.

Power Classification	2F
Designer	Webb
Company	LNWR
Driving wheel	4ft 5½ ins
Boiler pressure	150PSI
Cylinders	Inside 17" x 24"
Tractive Effort	16530lbf
Valve gear	Stephenson (slide valves)

Built specifically for freight work the first of this successful class of engines was withdrawn in 1903. BR continued withdrawals in 1948 with only 10 surviving to 1952. The last few remaining members of the class were employed at Crewe Works as shunters.

Webb 17" Coal Engine, seen in photographic grey primer prior to final painting at Crewe Works in 1873. LNWR Society

LNWR Bowen Cooke 'George the Fifth' 3P 4-4-0, 1910 to 1915.
90 engines Crewe built.
2 engines given BR Nos, 58011 and 58012.
None preserved.

Power Classification	3P
Designer	Bowen Cook
Company	LNWR
Driving wheel	6ft 9ins
Boiler pressure	180PSI superheated
Cylinders	Inside 20½" x 26"
Tractive Effort	20640lbf
Valve gear	Joy (piston valves)

Bowen Cooke was Whale's succesor and he designed his 'George the Fifth' class to be a superheated version of the 'Precursor' class. Crewe Works built 90 of the class which were numbered 5320-5409 in the early LMS system. The superheated engines were originally built with round topped fireboxes which over time were changed for Belpaire superheated boilers. Withdrawals began in 1935 and only 3 of the class came into BR stock. Of those LMS No 25321 LORD LOCH was scrapped in February 1948 without being allocated a BR number, 25350 became BR No 58011 INDIA and 25373 PTARMIGAN became BR 58012.

Top. Bowen Cooke 'George the Fifth' 3P 4-4-0 LNWR No 1489 WOLFHOUND at Crewe circa 1914. Withdrawn as LMS No 5341 March 1936. LNWR Society.

Bottom. Bowen Cooke 'George the Fifth' 3P 4-4-0 LNWR No 2664 QUEEN MARY at Manchester London Road. This engine built as a 'Queen Mary' class in 1910 and converted in 1913. Withdrawn as LMS No 5329 March 1936. LNWR Society.

LNWR Whale 'Precursor' 3P 4-4-0 1904 to 1907.
130 engines Crewe built.
1 allocated BR No 58010 SIROCCO.
None preserved.

Power Classification	4P
Designer	Whale
Company	LNWR
Driving wheel	6ft 9ins
Boiler pressure	180PSI superheated
Cylinders	Inside 20½" x 26"
Tractive Effort	20640lbf
Valve gear	Joy (piston valves)

The 'Precursor' class 4-4-0 design was considered to be the most efficient passenger engine ever owned and operated by the LNWR. A total of 130 engines were built at Crewe, LMS number system 5187-5319 with only one of the class becoming BR stock. The 4-4-0s were originally built with un-superheated round topped boilers however over time most of the class were rebuilt with superheated Belpaire boilers. The first engine was withdrawn in 1927 and LMS No 25297 was the only engine taken into BR stock (the last surviving member of the class) but the locomotive never carried its allocated number.

Top.   LNWR Whale 'Precursor' 3P 4-4-0 LNWR No 1117 VANDAL is seen when almost new at Shrewsbury depot circa1905. Withdrawn as LMS No 5194 during October 1931. F Moore/P.W. Pilcher.

Bottom. LNWR Whale 'Precursor' 3P 4-4-0 LMS No 5300 HYDRA is seen at Rugby depot during 1934. Withdrawn as LMS Duplicate No 25300 in June 1940. Rail Photoprints Collection.

LNWR Webb 'Coal Tank' 0-6-2T 1881 to 1897.
300 engines Crewe built.
291 to LMS stock, LMS Nos 7550 to 7841.
64 locomotives to BR stock, 6 withdrawn before numbers allocated. 58 engines 58880-58937.
1 preserved LNWR No 1054 (BR 58926)

LNWR COAL TANK

Power Classification	2F
Designer	Webb
Company	LNWR
Driving wheel	4ft 5½ins
Boiler pressure	150PSI
Cylinders	Inside 17" x 24"
Tractive Effort	16530lbf
Valve gear	Stephenson (slide valves)

LNWR Webb 'Coal Tank' No 848 seen at Crewe in photographic grey primer, circa 1888. Wikipedia.

This class of Webb designed 0-6-2T locomotives were effectively a tank engine version of his successful 0-6-0 'Coal Engines'. Crewe built 300 of the 'Coal Tanks' and they were allocated to LNWR depots all over the system. The engines were described as being reliable and versatile regularly working passenger trains. Later in their working lives some were fitted for 'push-pull' working. The first of the class was withdrawn in 1921 and none survived beyond 1958. The preserved engine LNWR No 1054 became LMS 7799 and then BR 58926, it was withdrawn during January 1958 from Shrewsbury (84G). In the present day, restored to working order, the 'Coal Tank' is often associated with the Keighley & Worth Valley Railway (K&WVR).

LNWR 'Coal Tank' 2F 0-6-2T LMS No 7722 is seen with a '6-wheel picnic saloon' outside Huddersfield station. This locomotive was withdrawn prior to 1948. LNWR Society.

Preserved 0-6-2T 'Webb Coal Tank' LNWR No 1054 is seen at The Keighley & Worth Valley Railway during the summer of 1986. Keith Langston.

Preserved 0-6-2T LNWR Webb 'Coal Tank' class No 1054 is seen whilst giving brake van rides on the demonstration line at the former Dinting Railway Centre, Higher Dinting, Sunday 3rd October 1982. David Ingham.

British Railways 'Standard Design Group'

The strain of maintaining railway services during the Second World War took a heavy toll of Britain's steam locomotives.

On 1 January 1948 the British Transport Commission (BTC) was formed by an act of parliament, the act placed all of the existing railway companies and their assets under the control of one organisation. Thus British Railways (BR) came into being. The railway infrastructure in general had suffered badly during the war years from the obvious pressure none more so than the steam locomotive operating departments.

A large percentage of engines were in poor condition due to their age, the lack of planned maintenance and the severe wartime working conditions. The re-creation of an effective railway system was essential if the countries infrastructure and manufacturing base were to be rebuilt.

Although plans to eventually replace steam power were already in place the BTC/BR boards decided in the meantime to design and build a new series of easily maintained steam locomotives, ideally with common design features.

A new organisation called the Standard Design Group was created and given a clear mandate to solve the operational problems caused by the shortages of effective steam motive power. The person chosen to head that organisation was Crewe Works man Robert Arthur Riddles. He liaised with a committee of Regional Chief Draughtsmen, chaired by Ernest Stewart Cox. There were 12 classes of 'BR Standard Locomotives' ranging from super powerful express and freight engines to suburban tank locomotives (999 locomotives in total). The building of the 'new' era locomotives took place at Crewe, Brighton, Darlington, Derby, Doncaster, Horwich and Swindon. Crewe Works built 3 classes exclusively, 'Britannia Pacific', 'Duke of Gloucester Pacific' and 'Clan Pacific', and they shared the production of a further 2 classes, the '84000 Class 2' 2-6-2T with Darlington and the '9F' 2-10-0 freight locomotive with Swindon. It should be noted that Crewe was the only works to build the 'Franco-Crosti' boilered variant of the '9F' class.

Robert Arthur Riddles C.B.E. (1892-1983)

Riddles served a long career in the railway industry which started in 1909 when he became a premium apprentice with the LNWR. He is regarded as being one of the greatest Crewe trained men. Although he never held the highest executive position at the works he went on to become Vice President of the LMSR (1944). He served in WWI with the Royal Engineers (France) where he was badly wounded. After that war he became the 'bricks and mortar' man at Crewe overseeing the building of several structures including the New Erecting Shop. In 1933 Riddles moved to Euston becoming 'Locomotive Assistant' to the new CME William Stanier, two years later he attained the post of 'Principle Assistant'. Amongst his many achievements was the design of the highly successful WD 2-8-0 and 2-10-0 Austerity locomotives. He retired in 1953.

Preserved BR Standard 'Britannia' Pacific No 70000 BRITANNIA is seen after the fitting of air brake equipment by LNWR Heritage Co Ltd. The pump/compressor etc can be seen neatly fitted onto the buffer beam and inside the smoke deflector. The associated air storage tanks are located under the engines tender. This locomotive was withdrawn from Newton Heath (9D) during May 1966. Keith Langston.

Standard 'Britannia' 7P6F 4-6-2,
BR 70000 to 70054 1951 to 1954.
55 engines Crewe built, 54 named.

CREWE   **BR**

2 preserved Nos 70000 BRITANNIA, 70013 OLIVER CROMWELL.

STANDARD PACIFIC

Power Classification	7P6F/7MT
Designer Riddles	British Railways
Designed Derby - Built Crewe	
Driving wheel	6ft 2ins
Boiler pressure	250PSI Superheated
Cylinders	Outside 20" x 28"
Tractive Effort	32160lbf
Valve gear	Walschaert (piston valves)

Standard 'Britannia' Pacific No 70011 HOTSPUR seen in the works on 16/01/1966. This locomotive was withdrawn from Carlisle Kingmoor (12A) 23/12/1967. Brian Robbins/Rail Photoprints

The new 4-6-2 locomotives designed in the Derby drawing office of BR were built exclusively at Crewe Works. BR Standard 'Britannia Class' were the first of 'Riddles' Standard type locomotives to enter service. Locomotive No 70000 BRITANNIA was completed at Crewe Works in January 1951, and was number one of the eventual 999 BR Standard locomotives built. The Britannia boiler was a superheated BR1 type, fitted with a self cleaning smoke box (SC), rocking grate and self emptying ash pan. To complete the new design, roller bearings and a tender cab were incorporated. The engines were fitted with a single blastpipe and chimney and Riddles choose two 20in x 28in cylinders, which he judged would reduce maintenance time and cost. The class constantly proved capable of maintaining express timings, often with heavily loaded trains equalling performances more normally associated with multi-cylinder 'Class 8' locomotives. After initial teething problems were solved the Britannia locomotives were hailed as capable high speed mixed traffic engines. With an axle loading of slightly over 20 tons they also enjoyed wide route availability. The Britannia class continued to work into 1964, but all were withdrawn before the end of BR steam.

BR Standard 'Britannia' '7MT' 4-6-2 No 70002 GEOFFREY CHAUCER passing Kirkstall with a train of mixed vans on 20/08/1966. This locomotive was withdrawn from Carlisle Kingmoor (12A) in January 1967. Mike Stokes/Keith Langston Collection.

BR Standard 'Britannia' 7MT' 4-6-2 No 70009 ALFRED THE GREAT, just one month old at Crewe North (5A), 30/06/1951. This locomotive was withdrawn from Carlisle Kingmoor (12A) in January 1967. Michael Bentley Collection.

BR Standard 'Britannia' 7MT' 4-6-2 No 70010 OWEN GLENDOWER at Sheffield Victoria station, LNER 'B1' No 61150 waits in the platform, circa 1959. The '9F' locomotive was withdrawn from Carlisle Kingmoor (12A), September 1967. www.railphotoprints.co.uk.

BR Standard 'Britannia' 7MT' 4-6-2 No 70010 OWEN GLENDOWER (already minus nameplates) is seen in the shed yard as viewed from the top of the coaling tower at Patricroft MPD in 1965. Jim Carter/Rail Photoprints Collection.

BR Standard 'Britannia' 7MT' 4-6-2 No 70013 OLIVER CROMWELL (later preserved) is seen with a local service near Wilmslow as part of the locomotives running in program prior to allocation, no shed plate carried, on 31/06/1951. This locomotive was withdrawn from Carnforth (10A) on 17/08/1968. Gordon Colas Trust/Mike Bentley Collection.

BR Standard 'Britannia' 7MT' 4-6-2 No 70015 APOLLO. Almost the end of its working life! Note that the BR style smokebox number plate, nameplates and shed code plate have all been replaced by replicas. The locomotive is being serviced at Stockport Edgeley in May 1967. This locomotive was withdrawn from Carlisle Kingmoor (12A) on 05/08/1967. Mike Stokes/Keith Langston Collection.

BR Standard 'Britannia' 7MT' 4-6-2 No 70020 MERCURY is turned at York MPD in preparation for heading the return 'Home Counties Railway Society' special to London Kings Cross on 04/10/1964. This locomotive was withdrawn from Carlisle Kingmoor (12A) in January 1967. RPC www.railphotoprints.co.uk.

BR Standard 'Britannia' 7MT' 4-6-2 No 70026 POLAR STAR seen near Chippenham with a local service forming part of the running in programme after attention at Swindon Works. This locomotive was withdrawn from Stockport Edgeley (9B) during January 1967. RPC www.railphotoprints.co.uk.

BR Standard 'Britannia' 7MT' 4-6-2 No 70030 WILLIAM WORDSWORTH is pictured at Willesden shed on 08/11/1964. It appears that locomotive cleaning was no longer top of the work schedule at 1A. However this engine continued in service for another 18 months before being withdrawn from Carlisle Upperby (12B) during June 1966. RPC www.railphotoprints.co.uk.

BR Standard 'Britannia' 7MT' 4-6-2 No 70034 THOMAS HARDY is seen at Crewe South MPD whilst being serviced in April 1963. The exterior condition of this engine is good compared with 70030 as seen above. This locomotive was withdrawn from Carlisle Kingmoor (12A) during May 1967. RPC www.railphotoprints.co.uk.

BR Standard 'Britannia' 7MT' 4-6-2 No 70039 SIR CHRISTOPHER WREN seen on the turntable at Kings Cross MPD on 22/02/1962. This locomotive was withdrawn from Carlisle Kingmoor (12A) during September 1967. *Rail Photoprints Collection.*

BR Standard 'Britannia' 7MT' 4-6-2 No 70040 CLIVE OF INDIA is pictured on shed at Patricroft MPD in March 1965. This locomotive was withdrawn from Carlisle Kingmoor (12A) during April 1967. *Jim Carter/Rail Photoprints Collection.*

BR Standard 'Britannia' 7MT' 4-6-2 No 70046 ANZAC (minus nameplates) is seen at Willesden MPD on 11/08/1963. This locomotive was withdrawn from Carlisle Kingmoor (12A) during July 1967. Rail Photoprints Collection.

BR Standard 'Britannia' 7MT' 4-6-2 No 70051 FIRTH OF FORTH is seen on the turntable at Patricroft during April 1963. Part of the Britannia class design specification called for the engines wheel base to be 58ft 3in, and therefore a size that would fit the majority of British Railways mainline turntables which were at least 60ft. This locomotive was withdrawn from Carlisle Kingmoor (12A) during December 1967. Jim Carter/Rail Photoprints.

BR Standard 'Britannia' 7MT' 4-6-2 No 70054 DORNOCK FIRTH seen at Slateford Junction with a 4 coach local service circa 1959. This locomotive was withdrawn from Carlisle Kingmoor (12A) during November 1966. David Anderson.

Preserved BR Standard 'Britannia' 7MT' 4-6-2 No 70000 BRITANNIA seen at LNWR Heritage, Crewe carriage shed site, in green lined livery with a white cab roof on 16/01/2012. Keith Langston.

Preserved BR Standard 'Britannia' 7MT' 4-6-2 No 70013 OLIVER CROMWELL seen at Carnforth (West Coast Railways) after hauling a mainline charter on 19/08/2008. This locomotive was withdrawn from Carnforth (10A) during August 1968. Fred Kerr.

BR Standard 'Duke of Gloucester' 8P 4-6-2, 1954.
1 engine Crewe built, named.

Engine preserved, 71000 DUKE OF GLOUCESTER.

BR Standard '8P' Pacific No 71000 DUKE OF GLOUCESTER, just arrived at Crewe station with the down Mid-Day Scot on 01/06/1956. Peter Kerslake.

British Caprotti

Rotary Cam Poppet Valve Gear

Power Classification	8P
Designer Riddles	British Railways
Designed Derby	Built Crewe
Driving wheel	6ft 2ins
Boiler pressure	250PSI Superheated
Cylinders 3	18" x 28"
Tractive Effort	39080lbf
Valve gear	British Caprotti

The fact that the BR locomotive got built at all is in itself an interesting piece of railway history. It was built as a direct result of the Harrow & Wealdstone rail crash which took place on 08/10/1952. During that tragic event two of the locomotives involved were damaged beyond repair, one of which was Stanier 'Princess Royal' class engine BR No 46202 PRINCESS ANNE. Riddles seized the loss of the Pacific to introduce the prototype of his Standard Design '8P' Pacific. During the planning stage of the new BR Standard locomotives, a provision was made to eventually include a large express passenger locomotive in the '8P' power category. Some in the higher echelons of BR no doubt expected that design to be a modified 'Coronation/Duchess' class 4-cylinder type. However the double chimney engine which emerged from the works was anything but, it was a 3-cylinder Pacific with an easily discernible difference. That being the inclusion of British Caprotti* poppet valve gear, a system which operated the engines valves via a series of shafts attached to the centre driving wheels. Riddles and his team anticipated that the new prototype '8P' would be the forerunner of a larger class of such engines. In the event it was the only one built. In service the engine fell short of the builders expectations, and No 71000 gained the reputation of being a poor steamer. The locomotive frequently worked on the WCML noticeably with the Euston-Crewe-Euston section of the 'Mid-Day Scot', but in its later days it was rostered to the less demanding North Wales Coast route. The 'Dukes' final rostered train was in fact a Crewe-Holyhead-Crewe stopping train before it was arguably prematurely retired, in November 1962. However, the anticipated success of the design was realised many years later in preservation, when after a great deal of hard work 71000's skilful restoration team rebuilt the magnificent engine and succeeded in unleashing the full potential of the Riddles '8P'. The locomotive spent all of its BR working life allocated to Crewe North (5A) MPD.

*The system was originally developed by Associated Locomotive Equipment of Worcester. They improved an earlier Caprotti system and also added another exhaust cam, the two cam arrangement delivered variable exhaust. The last two Black Fives 44686/44687 were also fitted with the British Caprotti twin exhaust cam system.

A magnificent sight! Preserved and restored Standard '8P' Pacific No 71000 DUKE OF GLOUCESTER visited Crewe Works in July 2004 in order to receive a new livery. The locomotive is pictured after the completion of that work and adjacent to the erecting shop in which it was built in 1954.
Keith Langston

Preserved Standard '8P' Pacific No 71000 DUKE OF GLOUCESTER is seen at Parton on the Cumbrian Coast with HF Railtours 'The Cumbrian Coast Explorer' on 02/07/2010. The steam at the rear of the tender indicates that the locomotives coal pusher is in use. Fred Kerr.

BR Standard 'Clan' 6P5F 4-6-2 1951/52.
72000 to 72009.

10 engines Crewe built, all named. None preserved.

BR Scottish Region

Pacific

Power Classification	6P5F/6MT
Designer Riddles	British Railways
Designed Derby	Built Crewe
Driving wheel	6ft 2ins
Boiler pressure	225PSI Superheated
Cylinders Outside	19 ½" x 28"
Tractive Effort	27520lbf
Valve gear	Walschaert (piston valves

BR Standard 'Clan' Pacific No 72000 CLAN BUCHANAN at Preston station in 1960. This locomotive was withdrawn from Polmadie (66A) on 29/12/1962. Keith Langston Collection. Tartan Buchanan Modern.

BR had plenty of names to choose from when naming the 'Clan' class. There were Highland and Lowland Clan names chosen from two distinct groups, those with chieftains and recognised by the heraldic Court of Lyon and those without chieftains thus being Armigerous Clans. The 'Clan' locomotives were intended to be a scaled down and lighter version of the 'Britannia' class with higher running plates, smaller diameter boilers and taller domes and chimneys. The engines smaller boilers were carried on the same chassis as the 'Britannia' class which with other differences resulted in reduced axle loading giving the engines the desired greater route availability. In service they were judged to be poor machines with steaming problems, lacking anything like the punch of the Britannias. All of the class were originally allocated to the BR Scottish Region but due to a lack of adhesive power did not as first intended, work regularly over the testing Highland Line. The class did however frequently worked Glasgow - Manchester and Liverpool express services. An order for a further 15 engines, 5 for BR Southern Region and a further 10 locomotives for the Scottish region was cancelled due to their poor performance and the quickening pace of dieselisation.

BR Standard 'Clan' Pacific No 72000 CLAN BUCHANAN seen between Crawford and Elvanfoot on the WCML, with a Glasgow-Blackpool holiday relief on 11 April 1960. David Anderson.

BR Standard 'Clan' Pacific No 72001 CLAN CAMERON is seen with a Glasgow-Manchester/Liverpool express approaching Crawford WCML . This locomotive was withdrawn from Polmadie (66A) in December 1962. David Anderson. **Tartan Cameron Modern.**

BR Standard 'Clan' Pacific No 72002 CLAN CAMPBELL seen on shed at Perth on 02/06/1956, having worked in from Glasgow Buchanan Street with a parcels train. This locomotive was withdrawn from Polmadie (66A) during December 1962. David Anderson. **Tartan Campbell Modern.**

BR Standard 'Clan' Pacific No 72003 CLAN FRASER seen on shed at Polmadie in 1956. This locomotive was withdrawn from Polmadie (66A) during December 1962. David Anderson. Tartan Fraser Modern.

BR Standard 'Clan' Pacific No 72004 CLAN MACDONALD seen making heavy weather of the climb up Beattock Bank with a 4 coach Carlisle-Glasgow Central stopping train 18/04/1959. This locomotive was withdrawn from Polmadie (66A) during December 1962. David Anderson. Tartan Macdonald Modern.

Macgregor

Modern

72005 at Carlisle Kingmoor in 1957.

Mackenzie

Modern

72006 at Carlisle Kingmoor in 1960.

Mackintosh

Modern

72007 on Beattock Bank in 1958.

Top. 72005 CLAN MACGREGOR was withdrawn from Carlisle Kingmoor (12A) during May 1965.

Middle. 72006 CLAN MACKENZIE was withdrawn from Carlisle Kingmoor (12A) during May 1966.

Bottom. 72007 CLAN MACKINTOSH was withdrawn from Carlisle Kingmoor (12A) during December 1965.

All images David Anderson.

BR Standard 'Clan' Pacific No 72008 CLAN MACLEOD is seen at Perth MPD in a rough outside condition in 1965. Only weeks away from the end of its working life. This locomotive was withdrawn from Carlisle Kingmoor (12A) during April 1966. David Anderson.

MacLeod

Modern

BR Standard 'Clan' Pacific No 72009 CLAN STEWART waits at Carstairs station with a Glasgow-Liverpool/Manchester express on 09/04/1955. This locomotive was withdrawn from Carlisle Kingmoor (12A) during August 1965. David Anderson.
Tartan Stewart Modern.

BR Standard 'Class 2 Tank' 2MT 2-6-2T 84000 to 84029 1953/57
20 engines Crewe built 84000-84019, 1953.
Darlington. 10 engines built 84020-84029, 1957.

None preserved.

BR Standard 'Class 2 Tank' No 84001 seen on shed at Llandudno Junction in May 1963. This locomotive was withdrawn from Llandudno Junction (6G) on 31/10/1964. Keith Langston Collection.

Push Pull fitted 2MT

Power Classification	2MT
Designer Riddles	British Railways
Designed Derby	Built Crewe
Driving wheel	5ft 0ins
Boiler pressure	200PSI Superheated
Cylinders Outside	16 ½" x 24"
Tractive Effort	18515lbf
Valve gear	Walschaert (piston valves)

Designated for light passenger work Ivatts LMS 2-6-2T passenger tank locomotive class was adapted for inclusion in the BR Standard range becoming the BR Standard '2MT' 2-6-2T class in the 84000 number series. Changes to the Ivatt design were minimal, they included boiler feed clack valves in place of LMS designed top feed and the addition of external regulator rodding. All of the class were originally fitted with vacuum control equipment to facilitate 'push pull' working, but later in their working lives that apparatus was removed. The 20 Crewe built engines were allocated to BR Midland Region depots with the 10 Darlington built examples being BR Southern Region allocated. Withdrawals began in 1963 with No 84012 being withdrawn after only 10 years and 1 month in traffic with the remaining members of the class being taken out of service before the end of 1965.

BR Standard 'Class 2 Tank' 2-6-2T No 84002 seen at Newport Pagnell having just arrived with the branch service from Wolverton, on 31/03/1962. This locomotive was withdrawn from Bletchley (1A) during April 1965. Note the BR jobs advertising poster on the station building. Neville Simms/Ranwell Collection/Rail Photoprints

BR Standard 'Class 2 Tank' 2-6-2T No 84003 is about to depart from Menai Bridge with the 10.55 Amlwch to Bangor service, on 22/08/1964. This locomotive was withdrawn from Llandudno Junction (6G) during October 1965. The service has used the now closed Amlwch Branch which was at the time of writing the subject of a long term reopening project, visit *www.Leinamlwch.co.uk*.
Ian Turnbull/Rail Photoprints.

BR Standard 'Class 2 Tank' 2-6-2T No 84006 is seen arriving at Wellingborough station, circa 1960. This locomotive was withdrawn from Leicester Midland (15A) during October 1965. RPC www.railphotoprints.co.uk.

BR Standard 'Class 2 Tank' 2-6-2T No 84008 takes water at Codnor Park while it heads the Stephenson Locomotive Society (SLS) Derbyshire & Notts railtour, 21/4/56, Derby Midland via Ripley, Wirksworth, Ilkeston, Heanor, Derby Friargate and Burton on Trent. This locomotive was withdrawn from Leicester Midland (15A) during October 1965. *Dave Cobbe Collection/Rail Photoprints*

BR Standard 'Class 2 Tank' 2-6-2T No 84014 seen at Stockport Edgeley MPD in December 1964. This locomotive was withdrawn from Eastleigh (70D) in December 1965. *Mike Stokes/Keith Langston Collection.*

BR Standard '9F' 2-10-0.
92000 to 92019 1954/55, 92030 to 92086 1954/56,
92097 to 92177 1956/58, 92221 to 92250 1958.
188* engines Crewe built.

BR Number series 92000-92019*, 92030-92250.
*Franco Crosti engines listed separately.
9 preserved 92134, 92203, 92207, 92212, 92214,
92219, 92220, 92240, 92245.

Swindon Works 92087-92096 1956/57, 92178-92220 1957/60.
1 locomotive named 92220 EVENING STAR.

BR '9F' 2-10-0 Flangeless Centre Drivers

Power Classification	9F
Designer Riddles	Designed at Brighton
Driving wheel	5ft 0ins
Boiler pressure	250PSI Superheated
Cylinders	Outside 20" x 28"
Tractive Effort	39670lbf
Valve gear	Walschaert (piston valves)
Giesl ejector instead of chimney 1 locomotive, 92250	
Franco Crosti locomotives shown separately. 92020/92029	

74 Standard '9F' locomotives were either built or later fitted with double chimneys, 3 engines Nos 92165/66/67 were fitted with mechanical stokers (later removed).

BR Standard '9F' 2-10-0 No 92002 is seen at Ebbw Junction circa 1958. This locomotive was withdrawn from Speke Junction (8C) on 11/11/1967.
Keith Langston Collection

Following WWII period the ex 'WD Austerities' and Stanier '8F' locomotives were the mainstay of BR's freight locomotives. There was an urgent need to replace/augment those ageing types with a new powerful BR Standard design of freight engine. After examining all possibilities Riddles and the Standard team settled on a 9F 2-10-0 design with 5 foot diameter driving wheels. The boiler design was new and designated as type BR9. The locomotive's boiler was pitched high on the frames to allow the grate to clear the rear coupled wheels. The grate itself was flat over the rear half and then sloped towards the front. A new design of non 'BR Standard' type regulator was used which operated by way of a sliding grid throttle in the steam dome which was activated by exterior rodding. The locomotives were designed to haul heavy freight trains at reasonably high speeds. In BR service the Standard '9F' engines also worked passenger trains at speeds reported to be between 80 and 90mph, a remarkable achievement for a small diameter driving wheel 2-10-0 design. The class had only a short service life, but long enough to prove their worth.

The 9F 2-10-0s were BR's most powerful locomotives, very reliable and also popular with engine crews. The 21ft 8in wheelbase could follow tight curves, helped by the centre pair of driving wheels being flangeless. In order to further widen route availability the designers kept the axle loading to only 15 tons 10cwt. The Standard Class 9s successfully hauled heavy mineral, coal and general freight trains. Westinghouse air pumps were fitted to 10 of the class in order to work the power operated wagon doors of mineral trains. The '9F' engines came into traffic between 1954 and 1960, with withdrawal commencing in 1964, 18 continued in traffic throughout 1967. The last 3 to be withdrawn in July 1968, were Nos 92077, 92160 and 92167.

BR Standard '9F' 2-10-0 No 92124 at Heaton Mersey in 1966. This locomotive was withdrawn from Warrington Dallam (8B) during December 1966.
Keith Langston Collection/Mike Stokes.

BR Standard '9F' 2-10-0 locomotives were regularly diagrammed to work oil tank trains out of Stanlow Refinery via the Helsby to Mouldsworth branch in order to join the former Cheshire Lines route. An unidentified member of the class is seen passing through Delamere Forest with a train for Trafford Park during the summer of 1966. Keith Langston.

BR Standard '9F' 2-10-0 No 92049 heads away from Chester with a train for the Minera Branch, Wrexham in September 1965. This locomotive was withdrawn from Speke Junction (8C) on 11/11/1967. Keith Langston.

BR Standard '9F' 2-10-0 No 92068 is seen at Annesley circa 1964. This locomotive was withdrawn from Derby (16C) during January 1966. Keith Langston Collection/Mike Stokes.

BR Standard '9F' 2-10-0 No 92109 seen working hard with a heavy freight near Swinton in 1958. This locomotive was withdrawn from Speke Junction (8C) during November 1967. Keith Langston Collection/Mike Stokes.

BR Standard '9F' 2-10-0 No 92112 is pictured passing through Delamere Forest with a train for the North West of England made up of oil tanks ex Stanlow Refinery in April 1967. Note the makeshift smoke box number plate. The DMU on the opposite track is operating a Manchester-Chester Cheshire Lines service. This locomotive was withdrawn from Speke Junction (8C) in November 1967. Keith Langston.

BR Standard '9F' 2-10-0 No 92133 is seen hard at work with a southbound freight at Moore on the WCML in April 1967. Another member of the class can be seen in the far distance with a northbound train of oil tanks. This locomotive was withdrawn from Birkenhead Mollington Street (8H) on 22/07/1967. Keith Langston.

BR Standard '9F' 2-10-0 No 92152 disturbs the peace and quiet of Delamere Forest on the former Cheshire Lines Chester-Manchester route as the locomotive heads north with a train of oil tanks from Stanlow Refinery in May 1957. This locomotive was withdrawn from Speke Junction (8C) during the week ending 11/11/1967. Keith Langston Collection.

Top. BR Standard '9F' 2-10-0 No 92224 seen near Wolvercot Junction, Oxford with an up freight in 1961. This locomotive withdrawn from Warrington Dallam (8B) in September 1967.
Middle. BR Standard '9F' 2-10-0 No 92239 seen near Basingstoke with a fully fitted freight in September 1961. This locomotive withdrawn from York North (50A) on 17/11/1966.
Bottom. BR Standard '9F' 2-10-0 No 92244 seen at Oxford shed on 16/06/1962. This locomotive withdrawn from Gloucester Horton Road (85B) on 31/12/1965. All Images David Anderson.

BR Standard '9F' 2-10-0, Franco-Crosti Boiler.
BR Number series 92020 to 92029 1955,
10 engines Crewe built. None preserved.

Franco-Crosti Boiler

Multi blast-pipe exhaust on the side of the locomotive.
Neville Sims/ Ranwell Collection/Rail Photoprints

Schematic diagram with single feedwater heater. Dan Crow.

BR Franco-Crosti Boiler.

In 1955 the Standard Design Team designed built and trialled '9F' locomotives with their version of the Franco-Crosti boiler system. The object was to save coal and also be able to burn lower quality coal. After approximately 3 years the performance data showed only minimum savings which were far outweighed by the problems encountered. In 1958 a decision was taken to temporarily take the 10 engines out of traffic.

Power Classification	9F Franco-Crosti
Rebuilds to single boiler	8F
Designer Riddles	Designed at Brighton
Driving wheel	5ft 0ins
Boiler pressure	250PSI Superheated
Cylinders	Outside 20" x 28"
Tractive Effort	39670lbf
Valve gear	Walschaert (piston valves)
Giesl ejector instead of chimney 1 locomotive, 92250	

The 'Standard 9F' Franco-Crosti locomotives had twin boilers, smaller fireboxes with the main boiler being a conventional tapered type situated on the frames, that was located above a smaller diameter pre-heat boiler located between the frames, each boiler had a separate smoke box door. This twin boiler system was developed in the 1930s by engineers Attilio Franco and Dr Piero Crosti. Basically the Franco–Crosti system uses the heat remaining in the locomotive's exhaust gases to preheat the main boilers water supply. This heat exchanger arrangement raised the temperature of the feed water, but did not produce steam. The energy in the exhaust gases is thus put to good use as the preheated water is then fed under pressure, into the conventional boiler. A Franco-Crosti engines conventional chimney is only used during the lighting up process. As the boiler starts to produce steam, the normal chimney is closed, with the hot gases being diverted through the feed water heater and then exiting the engine via a multiple blast pipe located on the side of the boiler. The anticipated fuel savings and performances were well below those expected by the designers. Furthermore the locomotives suffered from excessive corrosion problems in the tube water heaters, main boiler smokebox and chimney. Consequently in 1958 the BR Franco–Crosti engines were side-lined. Over the following couple of years the twin boilered engines were converted back to conventional operation, with the pre-heat boiler sealed off. The engines still retained their Crosti look and were not fitted with smoke box located smoke deflectors. After being converted back to single boiler operation the reconfigured engines were then designated 8F.

BR Standard Franco-Crosti '9F'2-10-0 No 92029 (the last Franco-Crosti to be built) is seen at Crewe Works 08/1955. This locomotive was withdrawn from Speke Junction (8C) in the week ending 11/11/1967.Neville Sims/Ranwell Collection/Rail Photoprints.

An unidentified reconfigured Franco-Crosti BR Standard '8F' is seen double heading with a Standard '9F' passing Delamere station on the former Cheshire Lines route. The secondary boiler can be seen blanked off behind the additional step on the buffer beam. The ex Stanlow Refinery oil train is heading for Trafford Park Manchester during April 1967. Keith Langston.

Franco-Crosti boilered locomotives are seen under construction at Crewe Works. The occasion was a public works visit during March 1955. The twin boiler arrangement can be clearly seen in this image. RPC www.railphotoprints.co.uk.

BR Standard Franco-Crosti '9F' 2-10-0 No 92020 when new at Crewe Works is seen in the company of sister engine No 92029. Interestingly the tender in primer, to the left, has *Paint Shop 1145* chalked on it. This locomotive was withdrawn from Birkenhead Mollington Street (8H) in the week ending 21/10/67. Neville Sims/Ranwell Collection/Rail Photoprints.

Reconfigured BR Standard Franco-Crosti '8F' 2-10-0 No 92021 seen on the coal road at a dilapidated coal stage at Kettering, during 1963. Note the coal stage worker looking on. This locomotive was rebuilt to conventional operation in 1960 and withdrawn from Speke Junction (8C) during the week ending 11/11/67. Rail Photoprints/John Day Collection.

BR Standard Franco-Crosti '9F' 2-10-0 Nos 92022 & 92023 stored awaiting conversion back to conventional steaming, seen at Wellingborough circa 1959, locomotive No 92022 reconfigured in 1963 and No 92023 in 1961. Both locomotives withdrawn from Speke Junction (8C) during the week ending 11/11/1967. RPC www.railphotoprints.co.uk..

BR Standard Franco-Crosti '9F' 2-10-0 Nos 92025 when seen at Crewe Works in 1958. By that time all the locomotives had been fitted with a boiler side smoke deflector to assist forward vision by clearing the exhaust. This locomotive was reconfigured in 1960 and withdrawn at Speke Junction (8C) at the end of December 1967. Neville Sims/ Ranwell Collection/Rail Photoprints.

BR Standard Franco-Crosti '9F' 2-10-0 No 92026 seen at Cricklewood in 1959 with rods removed, possibly prior to being despatched to Crewe Works, as it was reconfigured there in 1959. This locomotive was withdrawn from Speke Junction (8C) during November 1967. *Gordon Edgar Collection/Rail Photoprints.*

Reconfigured BR Standard Franco-Crosti '8F' 2-10-0 No 92027 is seen working hard on a freight near Bredbury on 30/09/1965. This locomotive was reconfigured in 1960 and withdrawn from Speke Junction (8C) in August 1967.
Paul Claxton/Rail Photoprints Collection.

Reconfigured Franco-Crosti BR Standard '8F' No 92026 is pictured on passenger duty with the afternoon Chester-Crewe leg of the LCGB 'Severn & Dee Railtour' on 26 February 1967. Keith Langston.

Reconfigured Franco-Crosti BR Standard '8F' No 92027 is seen at Warrington in August 1966. This locomotive was reconfigured in 1960 and withdrawn from Speke Junction (8C) during the week ending 05/08/1967. Keith Langston.

The Great Gathering - Webb Crewe Works Charity Fund.

Resplendent in LMS livery Stanier 'Princess Coronation/Duchess' 8P Pacific LMS No 6233 DUCHESS OF SUTHERLAND arrives at the Great Gathering. The mainline connection from the North Wales route, opposite the Crewe Electric depot, no longer exists and at the time of writing that former works area was under commercial development.
All Great Gathering images Keith Langston/Keith Langston Collection.

The highly successful Great Gathering event held on the works site over the weekend of 10/11 September 2005 was heralded as a triumph of organisation and a significant milestone in locomotive preservation. Preserved locomotives from all over the UK were brought together and assembled in an unrivalled 'works site' motive power display. The beneficiaries of the event were numerous local charities, organisations and preserved railways. The obvious winners were the railway enthusiasts, who in huge numbers flocked to the then Bombardier Crewe Works site.
The comprehensive collection of railway locomotives and associated exhibits attracted special charter trains, which were augmented by a shuttle bus service between the station and the works. For the most part the British weather behaved itself, the trains and buses ran on time and the available car parking coped with the influx of visitors to the famous Cheshire railway town, and what infrastructure remained of its legendary Locomotive Works. Those who were fortunate to be there will never forget the 'Gathering'. Whilst enjoying the occasion visitors were mindful that the size of the works had already decreased greatly from the halcyon days of railway locomotive building and aware that the shrinkage of the works could continue unabated. At the time of writing (2023) the famous site is still providing local people with railway related work under the ownership of Alstom, albeit a much smaller number of them. How long will Crewe have a railway engineering related industry? That is a question for the future.

The locomotives at the Great Gathering represented the former Big Four railway companies. In this celebration of Crewe Works only the locomotives originally built there are shown, when they were 'Back Home'.

Locomotive number(s) class and type	Date built	Withdrawn/Preservation
13268/2968/42968 LMS '5MT' 2-6-0	January 1934	31/12/1966 rescued 12/1973
5690/45690 LEANDER LMS 'Jubilee' 4-6-0	March 1936	17/03/1964 rescued 06/1972
6201/46201 PRINCESS ELIZABETH LMS 'Princess Royal' '8P' 4-6-2	November 1933	20/10/1962 rescued 02/1963
6203/46203 PRINCESS MARGARET ROSE LMS 'Princess Royal' 8P 4-6-2	July 1935	20/10/1962 rescued 05/1963*
6229/46229 DUCHESS OF HAMILTON LMS 'Princess Coronation/Duchess' 8P 4-6-2	September 1938	15/02/1964 rescued 05/1964*
6233/46233 DUCHESS OF SUTHERLAND LMS 'Princess Coronation/Duchess' 8P 4-6-2	July 1938	08/02/1964 rescued 10/1964*
46441 BR/LMS '2MT' 2-6-0	February 1950	15/04/1967 rescued 04/1967
8151/48151 LMS/WD '8F' 2-8-0	September 1942	13/01/1968 rescued 11/1975
71000 DUKE OF GLOUCESTER BR '8P' 4-6-2	May 1954	24/11/1962 rescued 04/1974
92134 BR '9F' 2-10-0	May 1957	10/12/1966 rescued 12/1980

* These locomotives were initially rescued by Sir Billy Butlin and displayed at his holiday camps, 6203 at Pwllheli, 6229 at Minehead, 6233 at Ayr.

The view in September 2005. Everything to the right of the steam locomotives is a housing development, gone are the traverser and workshops beyond. A tall boundary fence now follows the line of the nearside edge of the traverser, where the visitors were standing.

The former mainline access seen in August 2023 compared with the view of Stanier '8F' 2-8-0 No 48151 with B1 No 61264 and support coaches in tow arriving from the mainline in September 2005.

All 2023 images of the Alston Crewe Works site taken from adjacent public areas.

Locomotives on the Crewe Works Traverser

LMS 6201 PRINCESS ELIZABETH.

LMS/BR 46229 DUCHESS OF HAMILTON, everything to the front of the locomotive has been redeveloped as housing.

LMS 6233 DUCHESS OF SUTHERLAND. The famous traverser location is now covered by a housing development.

BR 46203 PRINCESS MARGARET ROSE with 'Royal Scot' headboard. Viewed looking towards the Gatehouse works entrance.

Four of Stanier's Finest. Nos LMS 6233, BR 46229, BR 46203 and LMS 6201. The tracks have been lifted however, the buildings to the right were part of the Alstom Crewe Works in 2023.

LMS No 5690 LEANDER with BR No 42968 posed in the evening sunshine. The red brick building glimpsed to the left of the locomotives was still the Works Stores in 2005. Now replaced by houses and a tall fence. The tracks leading down to the, still in situ Gatehouse, have been lifted.

BR No 46441, BR 48151 (withdrawn from nearby Cheshire town of Northwich) LMS 5690 and a little of No 2968. Looking towards the Gatehouse. The buildings beyond and to the right were a part of the Alstom Crewe Works facility in 2023. The tracks have been lifted.

'Ladies in Red'. Nos LMS 6201, BR 46203 and BR 46229. The line up was pictured in front of the Works Stores, the tracks and that building have been removed. Viewed from right hand back edge of the former traverser pit.

Locomotives on the Crewe Works Traverser

BR No 42968 the workshops have been demolished and replaced by a housing development.

BR No 46441.

BR No 48151 in the rain. The workshop buildings in front of the '8F', like the traverser have all now gone.

LMS No 6233 prepares to leave the works after the 'Gathering'. The building behind the locomotive were in 2023 part of Alstom Crewe Works.

Then under restoration BR '9F' No 92134 is posed in the erecting shop. This '9F' is the only single chimney version to be preserved. Since the 'Gathering' this locomotive has fully restored and returned to steam.

BR No 71000 DUKE OF GLOUSTER with headboard, 'The Great Briton'. Viewed along side the Works Stores looking towards the Gatehouse.

Locomotive No 71000 is coaled up on arrival. The 2023 image of the coaling point to the right is part of a public house car park. The gable end of the workshop has been renovated since 2005 but the locations are approximately the same.

This is a 2023 image of the Four Eagles PH, the rear of public house overlooks a development area approximately sited where No 48151 was seen shunting in 2005.

Early Sunday morning at the 'Gathering'.   The fence marks the location of the traverser, seen from the gatehouse.

A handsome Crewe trio, BR No 46441, BR No 42968 and LMS 5690 LEANDER.

Three cylinder locomotives, LMS 5690 LEANDER with Walschaert piston valve gear and BR 71000 DUKE OF GLOUCESTER with Caprotti Valve gear.

Lament to the End of Steam.

Catenaries and pantographs
May all be very well-
Appurtenances of an age
When electricity's the rage
And coal can go to Hell;

We'll try to put the blame on Fate
When current fails to alternate,
But railways held the highest stock
When engines ran on igneous rock.
Yes, Britain knew her greatest fame
In Old King Coal's refulgent reign.

Now linear Induction fills
The next progressive need,
And diesel turbines too are planned
To fuel us to the Promised Land,
At astronomic speed;

And when at length the L.M.R.
Goes wholly thermo-Nuclear
We'll shed a sad nostalgic tear
For those we Loved in Yesteryear.
Great Britain played her greatest role,
When locos. ran on best steam coal.

To be rendered soulfully to the tune of 'The Chancellors Song' from 'Iolanthe'.

This poignant missive dates from the 1960's and is believed to have been penned as BR No 70013 became the last steam locomotive to leave Crewe after attention (February 1967).
All efforts to trace the author(s) have been unsuccessful and so they respectfully remain 'Crewe Poet Anon'.

Oxley Depot 1965. Brian Robbins/Rail Photoprints

The poem was first published in *Crewe Locomotive Works and its Men by Brian Reed* 1982. David & Charles ISBN 0 7153 8228 4.

Bibliography

The following media sources are recommended reference points for this and other railway subjects.

Reed, Brian *Crewe Locomotive Works and its Men.* David & Charles Newton Abbot 1982.
Talbot, Edward *An Illustrated History of LNWR Engines.* Oxford Publishing Co 1985.
Longworth, Hugh *British Railways Steam Locomotives 1948-1968.* Oxford Publishing Co 2005.
Online BRDatabase https:www.brdatabase.info.
Online Six Bells Junction https://www.sixbellsjunction.co.uk .

Acknowledgements

Images supplied on a commercial basis by:-
Rail Photoprints Collection www.railphotoprints.zenfolio.com.   LNWR Society Images www.lnwrs.org.uk.

The author would also like to thank other accomplished archivists/photographers who have allowed their images to be used. They include David Anderson, Michael Bentley, Ben Brooksbank, Crewe Works Archive, John Firth, Alan Fozard, David Gibson, Ken Gray, Michael Halbert, Brian Jones, Dave Jones, Fred Kerr, Peter Kerslake, John Magnall Collection, Richard Metcalf Collection, Len Mills, Norman Preedy. John Pritchett, Mike Stokes Collection, Peter Skuce, Edward Talbot Collection, Malcolm Whittaker.

The 'Great Gathering' was made possible by the goodwill of the then site owners Bombardier Transportation. But without the team of willing helpers, representing the 'Webb Crewe Works Charity Fund,' it would not have been such an outstanding success. A group of volunteers and supporters are seen with the iconic DUKE OF GLOUCESTER locomotive, as the event came to a close.

About the author

Cheshire based Keith Langston is a widely published and highly respected photo journalist specialising in railway and other transport related subjects. His interest and vast knowledge of all things railway stem from being brought up in the North West of England as part of a large railway family. A lifelong railway enthusiast he has over the past 30 years worked as a contributing freelance writer and photographer for heritage sector journals whilst establishing himself as an accomplished author in his own right.